CAMBRIDGE LIBRARY COLLECTION

Books of enduring scholarly value

Earth Sciences

In the nineteenth century, geology emerged as a distinct academic discipline. It pointed the way towards the theory of evolution, as scientists including Gideon Mantell, Adam Sedgwick, Charles Lyell and Roderick Murchison began to use the evidence of minerals, rock formations and fossils to demonstrate that the earth was older by millions of years than the conventional, Bible-based wisdom had supposed. They argued convincingly that the climate, flora and fauna of the distant past could be deduced from geological evidence. Volcanic activity, the formation of mountains, and the action of glaciers and rivers, tides and ocean currents also became better understood. This series includes landmark publications by pioneers of the modern earth sciences, who advanced the scientific understanding of our planet and the processes by which it is constantly re-shaped.

Fossils of All Kinds

To the naturalist John Woodward (*c.*1665–1728), fossils (using the word in its widest sense of 'things dug from the earth') were 'much neglected, and left wholly to the Care and Treatment of Miners and meer Mechanicks'. He had built up a large personal collection of these samples of the Earth's petrified remains and spent much of his life developing a system for their classification, the results of which were published in this important illustrated work of 1728. A distinguished physician and a fellow of the Royal Society, Woodward wrote extensively on scientific topics, and had developed a theory that fossils were creatures destroyed in the flood described in the Bible. These ideas attracted critics and supporters in equal measure, but his contribution to techniques of fossil collection and classification were influential. In the present work, he devotes the early chapters to questions of description and classification, while the later sections contain some of his letters to his scientific contemporaries, including Isaac Newton.

Cambridge University Press has long been a pioneer in the reissuing of out-of-print titles from its own backlist, producing digital reprints of books that are still sought after by scholars and students but could not be reprinted economically using traditional technology. The Cambridge Library Collection extends this activity to a wider range of books which are still of importance to researchers and professionals, either for the source material they contain, or as landmarks in the history of their academic discipline.

Drawing from the world-renowned collections in the Cambridge University Library and other partner libraries, and guided by the advice of experts in each subject area, Cambridge University Press is using state-of-the-art scanning machines in its own Printing House to capture the content of each book selected for inclusion. The files are processed to give a consistently clear, crisp image, and the books finished to the high quality standard for which the Press is recognised around the world. The latest print-on-demand technology ensures that the books will remain available indefinitely, and that orders for single or multiple copies can quickly be supplied.

The Cambridge Library Collection brings back to life books of enduring scholarly value (including out-of-copyright works originally issued by other publishers) across a wide range of disciplines in the humanities and social sciences and in science and technology.

Fossils
of All Kinds

Digested into a Method,
Suitable to Their
Mutual Relation and Affinity

JOHN WOODWARD

CAMBRIDGE
UNIVERSITY PRESS

University Printing House, Cambridge, CB2 8BS, United Kingdom

Published in the United States of America by Cambridge University Press, New York

Cambridge University Press is part of the University of Cambridge.
It furthers the University's mission by disseminating knowledge in the pursuit of
education, learning and research at the highest international levels of excellence.

www.cambridge.org
Information on this title: www.cambridge.org/9781108068536

This edition first published 1728
This digitally printed version 2014

ISBN 978-1-108-06853-6 Paperback

FOSSILS

Of all KINDS,

DIGESTED into a

METHOD,

Suitable to their mutual

Relation and *Affinity*;

WITH

The NAMES by which they were known
to the ANTIENTS, and thofe by which they
are at this DAY known: And NOTES conducing to the fetting forth the NATURAL
HISTORY, and the main USES, of fome of
the moft confiderable of them.

AS ALSO

Several PAPERS tending to the further Advance-
ment of the Knowledge of MINERALS, of the
ORES of METALLS, and of all other SUB-
TERRANEOUS PRODUCTIONS.

By JOHN WOODWARD, M. D. late Pro-
feffor of Phyfick at GRESHAM COLLEGE,
Fellow of the College of PHYSICIANS, and the
ROYAL SOCIETY.

LONDON:
Printed for WILLIAM INNYS, at the Weft-
End of St. *Paul*'s. M.DCC.XXVIII.

THE
PREFACE,

Giving fome Account of the great
Plenty, Variety, and Excellence
of the fubterranean Productions and Riches of ENGLAND.

*ETALLS and Minerals are
allow'd on all Hands, to be of
that high Value, and of that
Ufe in fo many very important
Parts of human Life and Affairs, that
they merit, and juftly challenge our ut_
moft Study and Attention. The judicious
and intelligent Part of Mankind want
not a due Senfe of this, and that a great*

A 2   *Share*

*Share of our Wealth and Strength, our
Happiness and Security both at Home and
Abroad, depend very much upon them.
That a Confiderable Part of this Ifland,
I mean* Cornwall, *abounds in Tin, one of
the moft ufeful of them, has been known
from the earlieft Times. Nay, it has been
a chief Branch of our Trade, one of our
moft profitable Manufactures, and for
many Ages employ'd a multitude of
Hands, to great Advantage for themfelves and the Nation, that would otherwife have been wholly idle, in Want and
Diftrefs. The Lead of* England *is another
main Fund of our Riches. The Ore of
it is not only found here in great Plenty,
but 'tis kindly and well condition'd, melts
and obeys the Fire, and yields the Metall
in it with lefs Fuel, Trouble and Expence, than any of the Foreign Lead-
Ores that I know. Then, when feparated, 'tis better, fofter, more ductil, and
fit for Ufe, than that of all other Countries. Which does not arife from any Pe-
culiarity in the Metall; for the Lead of*
England, *and that of* Saxony, *the Gold
of* Japan, *and that of* Brazil, *the Silver*
                                    of

of Peru, *and that of* Norway; *to be
short, all Metall of the same Kind,
when reduc'd to an equal* Purity, *is
alike in every Respect, in what Country
soever it be got; but because the Spar,
and other extraneous Matter, incorporated with the* English *Lead in the Ore,
happens to be of such Nature and Disposition as to be wrought upon easily,
and freely to part from it. Nay, so
much* Iron *and* Copper *hath been discove-
red of late Years, and so many Ways of
working and extracting them newly found
out, that we have now vast Quantities of
our own to export and send Abroad, that
were wont heretofore to import them at
very considerable Charge. 'Tis but a few
Years since* Wad *or* Black-Lead *was found
out. Nor is there in all the whole Globe
besides, the like Plenty, or any of near the
Goodness and Worth of ours. The same may
be said of the Coal that we have without
Measure, and that is of such indispen-
sible Use and Necessity in almost all our
Affairs. Those large Masses of good
Load-Stone found on* Dartmoor, *the Antimony of* Cornwall, *the Manganese of*
Mandip,

Mandip, *and the Calamin, finer and better than any in the World besides, our Alum and Vitriol are further Instances of our Wealth under Ground. Many other Discoveries assuredly remain yet to be made, and Improvements in the Ways of working our subterranean Productions, managing them to better Advantage, and turning them to further Uses.* Multa egerunt qui ante nos fuerunt, fed non peregerunt, multum adhuc reftat operis, multumq; reftabit, nec ulli nato poft mille fæcula præfcindetur occafio aliquid adhuc adjiciendi, *Seneca. Those who have either little Capacity, and Command of Thought, or have it, and make little Use of it, will not be eafily brought to believe to how great Purposes Things feemingly very flight may be made ferviceable. For the Prefent, I will inftance only in* Fuller's-Earth ; *which* England *affords fo very good, and in Quantity fuperior to that of any Country befides. Those who are not rightly acquainted with the Uses of this, and should only look into the Pits of it, that are at* Wooborn, *and in feveral other Parts of the Kingdom, would be*
apt

ɪ

apt to slight and despise it ; and very probably to laugh at any Man who should take upon him to set forth how precious a Commodity it is ; tho' in Truth, it be a Thing of much higher Advantage, and brings in a greater Revenue to this Crown and Kingdom, *than the* Delves *of* Diamonds *in* Golconda, *the* Silver Mines *of* Potosi, *and the* Gold *of* Brasil, *bring into the* great Mogul, *the* King *of* Spain, *or* Portugal. *Those serve rather to reduce and impoverish the* People, *by rendring them proud and haughty, and consequently idle and vicious, than really to enrich and turn to their Benefit. Indeed their Neighbours wisely take the Advantage of their Sloth and Negligence ; and turn easily to their own* Profit, *what, want of Virtue and Industry in the original* Proprietors *let lye wholly unimployed and fruitless, while in their* Hands. *Our Ancestors were well aware of how great Benefit to the Nation* Fullers Earth *must needs prove. One main* Property *of it is to imbibe* Oil, Grease, *and all other like unctious Matter :* 'Tis *that* Property *that renders this Earth so useful*

useful in the cleanfing Woollen Cloth, *and freeing it from all thofe noifome and offenfive Impurities. Every Body con-verfant in rural Affairs, muft needs know how frequently* Tar *is of Neceffity imploy'd; as alfo Greafe and Tallow, in the Dif-eafes and Affections that Sheep are exter-nally fo frequently obnoxious to: And be-fides, their Wool cannot be work'd, fpun, or wore into Cloth, unlefs it be firft well oil'd and greas'd. Now, all this muft be taken out of it again, before it can be worn or turn'd to Ufe. Nor has there been any Thing ever yet found out fo fer-viceable to that End, as this Earth. And, as the* Fullers Earth *of* England *is got in great Plenty, fo it very much ex-ceeds any hitherto found out Abroad in Goodnefs. Which is the chief Reafon why the* Englifh *furpafs all other Nati-ons in the Woollen Manufacture ; and to preferve the Benefit of this to the Coun-try, and fecure it from the Ufurpation of Foreigners, the Exportation of* Englifh Fullers-Earth *is ftrictly prohibited by* Act of Parliament.

What

*What may serve further to incite our
Diligence and Curiosity is, that some late
Searches have shewn us many Things
besides those already pointed forth, that
were wont to be fetch'd from afar, nay
in Plenty, and much greater Perfection,
here at Home.* We have Demonstration
*of this in the many large stately Masses
of the blackest and most polite* Jeat, *discover'd so frequently on the Coasts of* York-
shire; *in the beautiful fine* Amber *of*
Suffolk, *and our other Shores. Then we
have* Jaspers, Cornelians, Agats, Mo-
choes, *and* Onyxes; *as also* Topazes,
and Amethysts, *as fine, if not so hard,
as the* Oriental. Diamonds *indeed we
have none; nor Rubies, with some others
of the Gemm-Kind. But, excepting
these, and* Cinnabar, *I know not any
Production of the Earth whatever, that
I have not found in this our native Country; such is the Præeminency of the Soil
of* England*! such its happy Fertility,
and Abundance in all Kinds of subterranean Treasure. Nor need we go far for
Proof of this, when my own Cabinets have*

now actually in them (to pass by the Extraneous, *which are in as great Numbers) above* 2800 Native Englih Follils,
*all different. So great a Number, got
together by the Indufry of one fingle
Man, involv'd all the while in Multipli-
city of other Bufinefs, cannot furely but
fhew that the Soil produces them in great
Abundance. Which will be made much
more apparent, whenever like Searches
are undertaken in Earnef by one that has
Leifure and Encouragement.*

   *But what crowns all is, a Man is here
fure, when with great Charge, Labour,
and Contrivance, he has once difcover'd
and obtain'd any Kind of this Treafure, to
hold and enjoy it. In other Countries, the
greatef Share falls to the Lord of the Soil,
or the Prince of the Country: And he
that fudies and drudges for it, enjoys the
leaft Part of it. This is a cruel Check
and Difcouragement to Search and Indufry! But happy* England *is fecure, and
wholly exempt from this; which is all
owing to the Virtue and Wifdom of our
Ancefors, and to the Excellence of thofe
Laws,*

*Laws, and that Conſtitution, which, at
the Expenſe of ſo much Blood and Trea-
ſure, they got eſtabliſh'd, and tranſmit-
ted down intire to us their Poſterity. By
which Means we continue a free People,
while not only our Neighbours all round,
but almoſt the whole World beſides, are
under a Tyranny of one Sort or other;
and ſubjeċted to the Luſt, Ambition,
Avarice, and Oppreſſion of thoſe who
ought to be Fathers of their Country, and
proteċt them in their natural Rights.
So ſacred a Tye, and high an Obligation
do we, who are ſo ſenſible, and thorowly
appriz'd of the Happineſs of theſe Laws,
and the Excellence of this Conſtitution,
lye under to guard both with the utmoſt
Zeal, Vigilance, and even Jealouſy;
to tranſmit them down intire and ſafe to
our Poſterity: To be true to that great
Truſt which our Forefathers have thus re-
poſed in us; and never part with, or give
up any the leaſt Particle of this ſo fair and
precious Jewel. If there be amongſt us any
of ſo clumſy a Frame, and ſo thorowly
hard-headed, that they cannot, or ſo cor-
rupt and blinded by other Intereſts, or ſo*

ſway'd

sway'd and byafs'd by wrong Maxims which they happen to be poffefs'd with, that they will not, be wrought upon by thefe Confiderations, they cannot fail of being effectually convinc'd, if they pleafe to look into what they will find fet in a much better Light by Polybius, *by* Livy, *by* Tacitus: *Or, if they pleafe to compare the Condition of the* Romans, *while a free State, with that, while under the continually incroaching Power of their Kings and Emperors; or of the* Athenians, *and the other* Grecian States, *while they were under the Protection and Encouragement of their own Laws, with that, when under their Conquerors: To obferve their then Grandeur, their Riches, and that all the more elegant and ufeful* Arts *and* Sciences *had their firft Rife, and the* Mufes *their Seat there; and compare that with their prefent Condition, their Meannefs, their Poverty, and even aftonifhing Ignorance, cannot furely hefitate one Moment in deciding to what it is that* Great Britain *owes all its Happinefs.*

Thefe

*These things rightly weigh'd, with se-
veral others that might well be offer'd;
had I not already too far transgress'd and
exceeded my Bounds; and the many noble
Products of* England *duly reflected on,
'twill hardly be possible for a Man to with-
hold himself from falling into the same
Transport and Passion for this Country,
that one of the greatest Wits of* Italy, *in
his Time,* Giovanni Cotta, *did for his.*

—— Qui Te noverit
Et non amarit protinus
Amore perditiſſimo,
Is, credo, ſeipſum non amat;
Caretque amandi Senſibus,
Et omnes odit Gratias.

*For those therefore that, thus taken
with these so useful, instructive and de-
lightful Studies, may, of their Virtue,
Good Sense, and Love to their Country,
be ambitious of facilitating them, and of
inlarging, and further displaying this so
beautiful and charming a Scene, I shall,
from my little Store, pick out such loose*
ſcat-

scatter'd Papers *that, I judge, may contribute something to their Light and Direction; delivering them in the Order
that follows.*

NUMBER I. Foffils digefted into a Method with Notes.

NUMBER II. In quâ, uno intuitu, con-
fpiciuntur omnis Generis Foffilia, juxta
ipfam naturæ Methodum, in Claffes
ordinata.

NUMBER III. A Letter to Sir *If. Newton,* fent along with the Method of
Foffils, giving an Account of the
Things needful and preparatory to the
drawing up fuch a Method. The Difficulties of it and its Ufes.

NUMBER IV. Letter to Sir *John Hof-
kyns* Baronet. The Study of Foffils
never hitherto reduced to Rule, nor
any Form of Art. The Writers, both
the Antients, and thofe of later Times,
have confounded Things buried in the
Earth, with the natural conftituent
Parts and Productions of it. Thefe
diftin-

diſtinguiſh'd, the Ranks of each adjuſted, and *Foſſils* divided into *Extraneous* and *Native*.

NUMB. V. Letter to the ſame. Of the
Cerauniæ, or *Stone-Weapons,* the *Magical Gemms,* and ſome other *artificial
Things,* antiently in uſe, imagin'd by
many Writers to be *natural,* with
Icons of ſeveral of thoſe in my Collection, brought from moſt Parts of
the known World.

NUMB. VI. Directions for regiſtring of
the *Native Foſſils,* and compoſing an
inſtructive and uſeful Catalogue of
them.

NUMB. VII. Letter to Monſieur——
at *Neufchattel.* The Aſſiſtance that
this, and ſeveral other learned Men
have given to the carrying on the
Deſign of the *Natural Hiſtory of the
Earth.*

NUMB. VIII. To the ſame. Of the Origin, Nature and Conſtitution of the
Belemnites.

NUMB. IX. To the ſame. Of the Coralloids, digg'd up at Land : The Nature and Origin of them.

NUM-

NUMB. X. Concerning Corall, Corallin, and other like Bodies form'd at Sea.

NUMB. XI. Brief Inftructions for making Obfervations and Collections ; and for compofing an *itinerant Regifter* of all things collected and obferv'd. Of Searches on the Surface of the Earth, upon Mountains; and in the Bowels of it, in Grottoes, Pits, Mines, and Quarries. Of the Water in Mines: Of Steams there, prefaging Changes of Weather: Of Damps, and other Meteors there. Of the Fogs, Mifts, or Clouds, that hover over high Mountains before Change of Weather. Of the Peat-Marfhes : And of the Trees, and other Things found buried in them.

NUMB. XII. An *Addition* to the fecond Part of the *Effay towards a natural Hiftory of the Earth*.

NUMB. XIII. A Mineral Dictionary; or an alphabetical Index of the Names of all Kinds of Foffils, referring to the Pages of this Work, wherein each is explain'd.

INDEX of Things *occafionally* treated of in thefe Papers.

A Me-

TABULA

In quâ, uno intuitu, conspiciuntur omnis generis FOSSILIA, juxta Naturæ METHODUM, in CLASSES ordinata.

FOSSILIA sunt.

1° Opaca, insipida, *friabilia*; in *aqua solubilia*: flammam non concipientia; TERRÆ; ad *Tactum*
  - *leves*, & quasi sebaceæ; quæ *Linguæ*, si illi admoveantur,
    - *adhærent.* CIMOLIA-PURPURASCENS. CIMOLIA-ALBA. ARGILLA. TERRA-SAMIA. TERRA-LEMMIA, tam RUBRA, quam ALBA. BOLUS ARMENA. KILLOIA. RUBRICA molliuscula.
    - *Non adhærent.* STEATITES. MOROCHITES. GALAXIA, seu LEUCOGRAPHIS.
  - *Scabra* & *Siccæ.* TERRA-VIRIDIS. TERRA CÆRULEA. RUBRICA duriuscula. TERRA-TRIPOLITANA. KILLOIA duriuscula. TERRA-CARIOSA. TERRA-MELITENSIS. TERRA-SINENSIS, e quâ Vasa porcellana dicta; sunt OCHRA. TERRA-FLAVESCENS. UMBRIA. CRETA. STEINOMARGA. *Geo. Agricola*, quæ est *Lac Lunæ Ol. Worm.* TERRA-NIGELLA, vegetabilis, Dædala. LUTUM. MARGA. S. *Terra rubella*, Zoica, Adamica. TERRA-MISCELLA.
    - *Appendix.* GLAREA, S. SABULUM. ARENA.

2° Insipida, *dura*, non ductilia, *nec in aqua solubilia*; LAPIDES; qui mole sunt
  - *Majores*, in Strata dispositi, *compositionis*
    - *laxioris*, ad tactum *scabri.* LAPIS-MOLARIS. Cos, tam GYRATILIS, quam PORTABILIS. SAXUM-ARENARIUM. SAXUM-SCABRUM. SAXUM SECTILE. SAXUM-CALCARIUM. SMIRIS.
    - *spissioris*, ad tactum *leves*, quiq; attritu *aliquatenus politi* fiunt. LAPIS-FISSILIS. LAPIS-LYDIUS. COS-OLEARIA. COTICULA.
    - *duræ* & *compactæ* adeo, ut ad *Nitorem* poliri possint. ALABASTRITES. MARMOR, colorum variorum. OPHITES. PORPHYRITES. GRANITA.
  - *Minores*; *Marmore.*
    - *Non duriores*:
      - *Figurâ* & *Constitutionis* incertæ & *indeterminatæ.* ROTULÆ-LAPIDEÆ. GLOBULI-LAPIDEI. LAPIDES-BORBORI. SCHERRI-LAPIDEI.
      - *Figurâ* extùs variæ & *interiùs, Constitutionis* verò internæ *determinatæ* & regularis:
        - Compositi, e *Fibris* parallelis, quæ in horum plerisq; *flexiles* sunt, & vi *elasticâ* præditæ. GYPSUM-STRIATUM, Anglicum. AMIANTHUS sive ASBESTOS. ALUMEN-PLUMOSUM.
        - Compositi e *Laminis* præsertim planis & parallelis, quæ *flexiles* sunt & vi *elasticâ* præditæ. TALCUM. MICA G. Agricolæ, argentea seu alba, uti ac aurea, & nigra.
        - Qui, interpositione laminarum e *Materiâ* ad *Fluores* dictos potissimùm accedente constansstituuntur *dividuntur in Tabes*, seu partes angulares, pentagonas, seu hexagonas, aut alias cujusvis figuræ angularis. LUDUS HELMONTII.
        - *Fistulosi*, ex *Tabulis* eâdem etiam *Materiâ* constantibus *compositi.* LAPIS-SYRINGOIDES.
        - Compositi è *Crustis* altera alteri *superinductis*;
          - *arctè cohærentibus* multâ intus Cavitate. BEZOAR-MINERALE.
          - *intus cavi*, cum *Materiâ* quadam inclusâ, Crustæ non adhærenti; sed mobili; *Solidâ*, & lapideâ; veteribus *Callimus* dicta. ÆTITES-SILICEUS. ÆTITES-OCHREO-FERREUS.
            - *laxâ*; uti Arena, Ochra, Creta, Terra, GEODES.
            - *liquidâ*, ENHYDROS.
      - *Figurâ* & *Constitutionis, certæ*, regularis & *determinatæ*, SELENITES. LAPIS-SPECULARIS. BELEMNITES, seu *Lapis Lyncis, Lyncurius* forte veterum. Corpora CORALLOIDEA-FOSSILIA, tam simplicia, quam ramosa. Lapides Coralloidibus Fossilibus affines STELECHITES. MYCPHITES. PORPITES. ASTROITES. LAPIS-FAVAGINOSUS. FUGORES, figuræ, CALCUM. CRASUM. STALACTITES. STALAGMITES. OSTEOCOLLA.
    - *duriores*;
      - *Opaci*;
        - *plerumq; unicolores.* LAPIS-NEPHRITICUS. MALACHITES. PRASITES. IASPIS-RUBENS ægyptius, *variorum* in eodem corpore *Colorum.* LAPIS-LAZULI, seu Cyaneus. HELIOTROPIUM. IASPIS.
      - *Semipellucidi*;
        - *versicolores*, prout vario situ luci objiciuntur, OCULUS CATI. OPALUS.
        - *coloribus* in subjecto *permanentioris* CATELLI, aliquot & SILICES COMPACTIORES & ELEGANTIORES. ACHATES. LAPIS CALCEDONIUS. ACHATES MOCHOENSIS. OCULUS-BELI. ONYX. SARDONYX. LAPIS-SARDIUS, seu CARNEOLUS vulgaris, CARNEOLUS-ALBUS: item LUTEUS, qui rarissimus. BERYLLUS GEM-MARIORUM, qui *Carneoli* species est magis pellucidi, & strraturæ rubentis.
      - *Pellucidi*:
        - *colorati*: TOPAZIUS Recentiorum, qui *Chrysolithes*, Veterum. HYACINTHUS, Gemmariorum. LAPIS-GRANATUS. RUBINUS *Rubrum.* Rubinus BALASSIUS. Rubinus SPINELLUS. CARBUNCULUS recentiorum, Rubinorum species rarissima. AMETHYSTUS. SAPPHIRUS, tam saturatè quam pallidè cæruleus, quæ SAPPHIRUS-AQUEA dicitur. Gemma Italis AQUA-MARINA dicta, quæ forte BERYLLUS *Plinii.* SMARACDUS. CHRYSOLITUS Recentiorum, qui *Topazius* Veterum.
        - *coloris expertes.* CRYSTALLUS. SAPPHIRUS ALBA. ADAMAS.

3° *Fissilia*, aliquatenus *pellucida*, linguam *pungentia*, in Aqua *solubilia*, ea autem evaporatâ, denuo *coalescentia* & in Figuras angulares se componentia. *Salia*. (§) SAL-FOSSILE, tam RUPEUM quam FONTANUM. Sal Cyrenaicum seu AMMONIACUM. TINCAL. *Perpetuum*, quod videtur CHRYSOCOLLA esse *Veterum.* NITRUM *Ægyptiorum veterum*, recentiorum NATRON, seu LATROS. NITRUM *recentiorum.* SAL-ACIDUM fossile, è quo, cum Materiâ *bituminosâ*, *cretaceâ*, vel *metallicâ*, coalescente, oriuntur SUL-PHUR, ALUMEN, & VITRIOLUM.

4° *Flammam* facile concipientia & *Oleum* præbentia, in *Aquâ non solubilia*; *Bitumina*;
  - *Liquida.* NAPTHA. PETROLEUM. OLEUM TERRÆ BARBADENSE.
  - *Solida.* BITUMEN. PISSASPHALTON. SUCCINUM. GAGATES. Lapis AMPELITES. LITHANTRAX.

5° *Metallis affinia*, quibusdam scilicet Metallorum Proprietatibus prædita, *Pondere* saltem & *Splendore*; *Mineralia*;
  - *Fluida*, ARGENTUM-VIVUM-NATIVUM.
  - *Solida*, *igne fusilia*, sed *non ductilia*, CINNABARIS. ARSENICUM aureum. ARSENICUM RUBRUM, seu SANDARACHA. PYRITES. MARCASITA. COBALTUM. LAPIS-CALAMINARIS. ANTIMONIUM. BISMUTHUM. SPELTRUM. NIGRICA-FABRILIS.

6° *Ponderosa*, *splendentia*, *solida fusilia*, & *ductilia*; *Metalla*. (§) AURUM, ARGENTUM. CUPRUM. FERRUM. STANNUM. PLUMBUM.
  - *Appendix ad Cap. de Ferro.* HÆMATITES, S. SCHISTOS, MAGNES, MAGNESIA, ferri plus minùs in se continent.

A
Methodical Diſtribution
O F
FOSSILS,
Of all Kinds, into their proper
CLASSES,

Viz. 1. *Earths,* 2. *Stones,* 3. *Salts,* 4. *Bi-*
tumens, 5. *Minerals,* 6. *Metalls.*

Claſs 1. EARTHS,

R Bodies opake, inſipid, and,
when dryed, friable, or conſiſting of Parts eaſy to ſeparate,
ſoluble in Water ; not diſpoſed
to burn, flame, or take fire.

B

C A-

CAPUT 1. Those that, to the Touch, have a Smoothness like that of unctuous Bodies.

MEMBR. 1. Such as, if applied to the Tongue, adhere to it. Fullers-Earth (1). Tobacco-Pipe-Clay (2). Potters-Clay (3). The Samian Bole, and the Lemnian, both the red and the white; Bole-Armeniac (4). The softer Killow (5), the softer *Ruddle*, or, as 'tis call'd in the North, Smitt (6).

MEM-

(1) *This is call'd by some Writers,* Cimolia purpurascens.

(2) Cimolia alba.

(3) Argilla.

(4) *These astringent Earths take their Names from* Samos, Lemnos, *and* Armenia, *the Countries from which we have them.*

(5) Killoia molliuscula. Killow *is found in* Lancashire, *and mentioned by* Dr. Merret *in his* Pinax. *'Tis of a blackish or deep* blue Colour, *and, doubtless, had its Name from* Kollow, *by which Name, in the North, the Smut, or Grime, on the Backs of Chimneys, is call'd*

(6) Rubrica molliuscula. *A sort of Earth of a dusky red Colour, found chiefly in* Iron Mines, *the finest in those of* Langron *in* Cumberland. *Some of this Earth contains as much* Iron *as to render it worth* smelting.

3

MEMBR. 2. Such as will not adhere to the Tongue. Soap-Earth (⁶*), French-Marking-Stone (⁷).

CAP. 2. Thoſe that, to the Touch, are dry, harſh, and rough. Terre Verte (⁸). Terre Bleue (⁹). The harder Ruddle (¹⁰). Tripoly (¹¹). The harder Killow (¹²), or *Marking-Stone.* Rotten Stone (¹³). Maltese-Earth (¹⁴). China Earth (¹⁵). of which the fine Earthen-ware of *China*

B 2   and

(6*) *Steatites.*

(7) *This probably is the* Mo ochites *of* Pliny : *and the* Moroéthus, Galaxia, *and* Leucographis *of* Dioſcorides. *It is unctuous to the Touch, as the former is, but harder and nearer approaching the Conſiſtence of Stone. The* French *call* it Craye de Brianſon.

(8) Terra Viridis. *This owes its Colour to a ſlight Admixture of* Copper.

(9) *So does the* Terre bleue, *which is no other than a light, looſe, friable Kind of* Lapis Armenus.

Terra cœrulea.

(10) Rubrica duriuſcula. *This owes its Colour to an Admixture of* Iron : *And as that is in greater or leſs proportion, the Body has a greater or leſs ſpecifick Gravity, and Conſiſtence, or Hardneſs.*

(11) Terra Tripolitana.

(12) Killoia duriuſcula. *This* Dr. Merret *calls* Lapis cœruleus ducendis Lineis idoneus.

(13) Terra cariofa.

(14) Terra Melitenfis.

(15) Terra Sinenfis.

and *Japon* is made.   Yellow-OCHRE ([16]).
GHALOLINA   ([17]).   UMBRE   ([18]).
CHALK ([19]).   The STEINOMARGA ([20]).
The BLACK-EARTH, every where obvi-
ous on the Surface of the Ground, which
we call MOULD([21]). Garden-Earth, or Un-
der-Turf-Earth.   Common CLAY ([22*]).
MARL ([22]). LOAM ([23]).

APPEN-

([16]) Ochra.

([17]) *Earth of a bright Gold Colour, found in the Kingdom of* Naples, *very fine, and much valued by Painters.*   Terra flavef-cens.

([18]) Umbria.

([19]) Creta.

([20]) Steinomarga Agri-colæ de Nat. Fof. *L.* 2. *p.* 578. Agarico minerale *Fer.* Imperati *Hift. Nat. L.* 5. *c.* 41. Lac. Lunæ Ol. Wormii Muf. *L.* 1. §. 1. *c.* 4. *This, when pure, is foft, light, and very white. 'Tis frequently found in Form of a white farinace-ous Powder, but fometimes* concreted into a Mafs, foft, fungous, and not unlike Agaric. *When there is a fmall Proportion of a fparry or arenaceous Matter in-corporated with it, it ren-ders it gritty and friable.*

([21]) Terra nigella ve-getabilis Dædala. *Concer-ning this fee the Introdu-ction to the* natural Hifto-ry of the Earth.

([22*]) Terra rubella, Zoica, Adamica. Lutum.

([22]) Marga.

([23]) *This is only* Marl or common Clay, *with a fmall Admixture of* Sand *in it.* Terra Mifcella.

APPENDIX to Claſs 1.

GRAVEL (a) and SAND (b).

THESE do not properly belong to this Place; yet, in compliance with the common Method of the Writers of Foſſils, I ſhall mention them here; and at leaſt point forth what they are.

(a) GRAVEL, *Glarea, Sabulum,* con-ſiſts of *Flints,* of all the uſual Sizes, and Colours; of the ſeveral ſorts of *Pebles;* ſometimes with a few *Pyritæ,* and other *Mineral Bodies,* confuſedly intermix'd; and common *Sand.*

(b) SAND, *Arena,* Ἄμμος, ⱷάμμος. Un-der this Title we have four Sorts of very different Bodies, *viz.*

1, *Extremely ſmall Pebles,* many of them white, ſeveral pellucid, ſome yel-low, red, and of other Colours. Theſe conſtitute the true, which is indeed our

common

common Sand; this being found in the *Gravel-pits* all over *England*, and particularly in thofe about *London*, in the Sand-Pits of *Hide-Park*, thofe about *Kenfington*, thofe near *Woolwich*, and upon *Blackheath*. Our Microfcopes fhew it to be only a *Congeries* of fuch fmall *Pebles*. The fame fort of *Sand* is alfo found on the *Shores* of the *Sea*, and *Rivers*; 'tis here commonly very clean and fine, the Waters ferving to wafh, clear, and free it from Earth, Clay, Mud, and other lighter Matter; and, by that Means, to bare and uncover the *Sand*, whenever the Earth there contains any in it.

2. *The Gritt of Stone*, or Matter, of that fort of which the *Strata* of Stone are compofed, found lying loofe. Part of this, by reafon of the Intermixture of Matter with it, that was earthy, lax, and incapable of Coalition, has not been confolidated, but lay ever loofe, and in the State in which it is now found. The reft is fuch as has by little and little moulder'd down after Frofts; and been beat off, from the *Strata*, by the Falls of Rain, or,
where

where it happens to be near them, by the Waves of the Sea, and Rivers. 'Tis found chiefly on the Sides, and at the Bottoms of Rocks ; and on the Shores of the Sea and Rivers.

3. *A brittle Shattery fort of Spar*, found, in Form of a white Sand, chiefly in the perpendicular Fiffures, amongft the Ores of Metalls.

4. *Small Fragments of Shells*, broken, and reduced into Form of Powder, by Means of Stones, and other ponderous hard Bodies, agitated by Tides and Storms. This is found in vaft Plenty on fome Shores, and is frequently made ufe of for the manuring of Land, by the Name of *Sea-Land.* See the *Reflettions concerning Vegetation, Philofoph. Tranfatt.* N°. 253.

Clafs 2. S T O N E S.

O R Bodies infipid, hard; not ductile, or malleable; nor foluble in Water.

C A-

CAPVT 1. Thofe which are found in great Maffes, and formed into *Strata.*

N. B. *The* Characteriftic *of the Bodies in* this Chapter, *I mean, their being* formed into Strata, *does not hold fo univerfally, but that there are fmall Deviations from it.* *Thus fometimes,* Marble *is found, not in* Strata, *but in the* perpendicular Fiffures *of them;* which Alabafter *likewife is, in fome Places, and indeed even a fine* Stony Matter, *as alfo an* Earthy, *e. g.* Umbre, *and* Ochre. *On the contrary,* Mineral *and* Metallic Matter, *found moft commonly in the* Fiffures, *is fometimes likewife found in the* Strata, *e. g.* Spar, Iron, Copper, *and the like.* *Nor can this be thought ftrange, to any one that rightly reflects upon the* Confufion *that thefe Bodies were in, after the* Diffolution *that befell them during the* Deluge ; *and upon the* Tranfitions *and* Removes *that are made by* Water *paffing the* Strata *into thofe* Fiffures. *Vid.* Nat. Hift. Earth. *Part* 2 *and* 4. *But this whole* *Affair*

Affair will be fet in a Light more clear, full, and diftinct, whenever the Catalogues *of my* Foffils, *both* Englifh *and* Foreign, *fhall come forth.*

MEMBR. 1. Such as are of a Compofition more lax, and a Grain more coarfe, or rough, to the Touch. Mill-Stone ([24]). Grind-Stone ([25]). Whet-Stone ([26]). Sand-Stone ([27]). Rag-Stone ([28]). Free-Stone ([29]). Flag-

C Stone

([24]) Lapis molaris.
([25]) Cos gyratilis.
([26]) Cos portabilis.
([27]) Saxum arenarium.
([28]) *So named from its breaking in a* ragged, *uncertain, irregular Manner.* Saxum Conftitutionis du-rioris, craffioris *Scabræ.*
([29]) *So named from its being of fuch a Conftitution as to be wrought and cut* freely, *in any Direction.* Saxum Sectioni in omnem Partem, & directionem, ex æquo cedens.

STONE (³⁰). LIME-STONE (³¹). *Polifh-ing Stone,* or EMERY (³²).

MEMBR. 2. Such as are com-monly of a clofer Compofition, and fome-what finer Grain, fo as to be more fmooth to the Touch, and in fome fmall Degree capable of a Polifh. SLATE (³³). TOUCH-STONE (³⁴). OIL-STONE (³⁵). The HONE (³⁶).

MEM-

(30) Saxum laminofum 'Tis call'd commonly Slate, merely becaufe 'tis us'd, and indeed very fitly, like Slate, for the covering of Houfes, particularly at Bath and in feveral Parts of the Weft. But it will not fplit, as Slate does, be-ing found form'd into what they call Flaggs, or thin Plates; which indeed are no other than fo many Strata. I have obferv'd of them, betwixt Caftle-ton and Workfworth, in the Peake of Derbyfhire, and in fome other Places, from the Thicknefs of Pa-per, thro' all Degrees to a very confiderable Bulk. They increafe defcending, the thickeft lying ever deepeft in the Earth. All the Strata of our Globe are compiled of terreftrial Matter fubfiding from the Water of the Deluge: and, when the Subfidence firft began, that Matter was in greateft Quantity; fo that the Strata that lye deepeft, muft of Courfe be the thickeft; and muft grow gradually thinner, afcending towards the Sur-face of the Earth, as the Water became more and more difengag'd of it.

(31) Saxum Calcari-um.

(32) Smiris.
(33) Lapis fiffilis.
(34) Lapis Lydius.
(35) Cos Olearia.
(36) Coticula.

MEMBR. 3. Such as are of a Conſtitution ſo hard and compact, and a Grain ſo fine, that they will readily take a bright Poliſh. ALABASTER (37). MARBLE (38) of divers Colours, both ſimple and mix'd, and found in ſeveral Countries, whence it has obtained ſeveral Names, which will be too tedious, and indeed of little Uſe to recite here. The OPHITES (39). PORPHYRY (40). The GRANITE (41) of the *Italian* Writers.

C 2   CAP.

(37) Alabaſtrites.
(38) Marmor.
(39) *The Ophites of the Moderns has a dusky greeniſh Ground, with Spots of a lighter Green, oblong, and uſually near ſquare. The* Ophites *of the Antients, was little if at all different, as appears from the Fragments of it ſtill remaining in Antient Works. Beſides,* Pliny's *Account agrees well with this. He calls it alſo* Memphites *from* Memphis *in* Egypt, *near which City 'twas got.* Plin. Nat. Hiſt. L. 36. c. 7.
(40) Porphyrites.

(41) Granita. *This is the* Syonites *and* Pyrrhopœcilus *of* Pliny, Nat. Hiſt. L. 36. c. 8. *which, according to his Intelligence, was got near* Syene *in* Thebais. *He obſerves, and indeed very rightly, that the* Egyptian *Obelisks are made of this.* V. M. Mercati de gli Obeliſchi di Roma, c. 2. p. 4. *It has been long a Doubt, amongſt the Learned, where ſo great a Quantity of* Porphyry *and* Granite, *as we ſee in the works of the Antients, yet extant, in* Syria, Phœnicia, Greece, *and* Italy, *was*

CAP. 2. Thofe which are found in fmaller Maffes.

MEMBR. 1. Such as do not exceed Marble in Hardnefs.

ARTICVLVS 1. That are both of a Figure and a Texture that is uncertain and undeterminate. Thofe call'd RUBBLE-STONES (⁴²). *Copple-Stones*,

was all digg'd up. But I have learned from the late Obfervations and Travels of Mr. H. Worfely, *and fince of Mr.* Tho. Shaw, Lett. Dec. 20. 1725. *and of fome other curious and intelligent Perfons, that there are many vaft* Strata, *and even whole Rocks, confifting intirely of thefe two Kinds of Marble in* Arabia Petræa. *Whence thefe might be eafily carried acrofs the* Red-Sea *into* Egypt; *and, by the Mediterranean, into* Phœnicia, Greece, *and* Italy.

(42) Rotulæ lapideæ. *The Water, at the latter End of the Deluge, depar-* *ting in Hurry, and with great Precipitation and Violence, bore with it, not only the loofer terreftrial Matter, but the Nodules and harder; nay, it tore up the very ftony* Strata, *broke them, rowl'd, and tumbled along the Pieces and Fragments frequently very far from the Places where they originally lay, rounded, fmoothed them, and brought them to Form of Nodules. They owe their Name,* Rubble, *to their being thus rubb'd and worn. Thefe we find, in fuch Countries where there is Stone, frequently in great Numbers, and of*
various

Stones, or BOWLDER-STONES (43).
CLAY-STONES (44). The *Stony Nodules*
found lodg'd in the *Strata,* and call'd by
the Workmen KNURS and KNOTS (45).

ARTIC.

various Sizes, in digging juſt within the Surface. But there are in many places, in Wales, *in* Cornwall, *and elſewhere, Maſſes of Stone, ſometimes to a vaſt Bulk. e. gr. of one, two, or more Tuns, thus torn up, and left at the Surface ; of which I intend a further Account in its Place.*

(43) Globuli lapidei.— *Theſe are found on the Shore : of the Sea and Rivers : are Lumps and Fragments of Stone or Marble, broke from the adjacent Cliffs, rounded by being* bowl'd, *and tumbled to and again by the Action of the Water. Whence they obtain'd the Name of* Bowlder Stones ; *they being form'd by an Action like that of a* Bowl, *and thereby reduc'd to the Shape of one. Neither the* Bowlders, *nor* Rubble-Stones, *are ever inveſted with an exterior ſtony* Cruſt *or*

Skin. *'Tis. plain, from Conſideration of the Manner of their Formation, they cannot. This is one Mark by which they are diſtinguiſh'd from* Flints, Pebles, *and the other* native Nodules, *that were form'd before the Subſidence of the Matter of the* Strata, *and are cover'd with ſuch a* Cruſt *or* Skin, *unleſs it have been worn off by their having been, ſince their Formation, likewiſe ſo agitated and worn.*

(44) Lapides borbori.

(45) Schirrhi lapidei, *From their being, as* Knots *in Timber, commonly harder than the reſt of the Maſs of the* Strata, *wherein they are found repoſited, whether that be of* Chiver, Slate, *or* Stone, *in each of which they are found uſually few in Number, of different Size, Subſtance and Shape, but commonly approaching a Globular.*

ARTIC. 2. That are *external-ly* of Figure various and uncertain; *but,* *internally*, of a Texture determinate and regular.

SECTIO 1. Thoſe which are compoſed of *Fibres*, which are parallel, and which, in moſt of them, are flexible, and elaſtick. ENGLISH TALC (⁴⁶), of which the coarſer Sort is call'd *Plaiſter*, or PARGET (⁴⁷), the finer, SPAAD (⁴⁸), EARTH-FLAX (⁴⁹), or *Salamander's Hair.*

SECT. 2. Thoſe which are compoſed of *Plates*, that are generally plain and parallel, and that are flexible and elaſtick. TALC (⁵⁰). *Cat-Silver,* or GLIMMER (⁵¹), of which there are three Sorts, the *Yellow* or *Golden,* the *White* or *Silvery,* and the *Black.*

SECT.

(46) Gypſum Striatum ſtius Straboni *L.*90 Geogr.
(47) Gypſum.
(48) Spatum.
(49) Amianthus, or
Asbeſtos. Lapis Caryſti-
(50) Talcum.
(51) Mica Geo. Agri-
colæ, aurea, argentea, ni-
gra.

SECT. 3. Thofe which, by the Interpofition of *Laminæ*, or Plates, confifting of a Talky Spar, are divided into Tali, or angular Parts, as Pentagons, Hexagons, or of fome other angular Figure. The Waxen-Vein (52) of Dr. *Grew. Catal. Muf. Soc. Reg. Lond.*

SECT. 4. Thofe which are fiftulous and compofed of *Pipes*, confifting of a like Talky Spar. The Piped-Waxen-Vein (53) of Dr. *Grew. Ibid.*

SECT. 5. Thofe which are compofed of *Crufts* including one another.

SVBDIVISIO 1. Having the Crufts adhering clofe to each other, ordinarily to the Center of the Body, without any Cavity within. Mineral Bezoar (54).

SVB-

(52) Ludus Helmontii.

(53) Lapis Syringoides.

(54) *Of the* Bezoar Minerale, *fee* P. Bocconne's Recherches & Obf. Nat. 8o.

SUBDIV. 2. Having a Cavity within, containing in it Matter, not adhering to the Cruſt, but looſe and moveable.

§ § §. 1. Solid and Stony, call'd by the Antients *Callimus.* The FLINTY-EAGLE-STONE (55). The OCHREOUS-EAGLE-STONE (56).

§ § §. 2. Lax; e. gr. Sand, Ochre, Chalk, Earth; the ELFS-EARTH-SCRIP (57).

§ § §. 3. Liquid; the FAIRY's-WATER-BOTTLE (58).

ARTIC. 3. That are of a certain, regular, and determinate Figure, and Conſtitution. The Rhomboidal SELE-

(55) Ætites Silicius.
(56) Ætites Ochreoferreus.
(57) Geodes. *There's one ſort of this found commonly among the clay us'd for making Tyles and Bricks; which the Work-* men call Race *or* Rance. *The* German *Mineraliſts give it the Name of* Erdmangen, *or* Earth-man.
(58) Enhydros. Ad motum, fluctuat intus in eo, veluti in ovis, liquor, *Plin.* xxxvii. 12.

SECT. 1. Thoſe which have Colours, changeable according to the different Poſition of the Stone to the Light. The CATS-EYE ([81]). The OPAL ([82]).

SECT. 2. Thoſe which have the Colours fix'd and permanent. The harder and finer PEBLES ([83]), and FLINTS ([84]). The AGATE ([85]). The CAL-

---

([81]) Oculus Cati. *This is of a gliſtering Grey, interchanged with a Straw Colour: And anſwers the Deſcription of the* Aſteria, *given by* Pliny. *The Ancients aſſign'd that Name only with Regard to the Brightneſs and ſhining of the Stone; without any Conſideration of Figure, which the Moderns ſeem only to have minded in their* Aſteria.

—Ἀστήριος, καλὸς, λίθος διατίς Ἀστήρ. Μαρμαίρων —Dionyſ. Περιηγ.

([82]) Opalus. *In this there is an interchangeable Mixture of Red, Green, Yellow, and Blue. We have this Stone uſually from* Germany. *It anſwers the Character of* Pliny xxvii. *6. and doubt-* leſs *is the ſame with the* Opal *of the Antients.*

([83]) Calculi. ⎱
([84]) Silices. ⎰*Some of theſe are, thro' the whole Inſide, of the ſame Colour, Black, Brown, Grey, White: others ſpotted, or lineated with various Colours. The* Germans *call our Flint* Hornſtein, *tho' Flint is moſt commonly found in Form of* Nodules: *But 'tis ſometimes found in thin ·Strata, when 'tis call'd* Chert.

([85]) Achates. *Agats are only Varieties of the* Flint Kind; *they have a grey horny Ground, clouded, lineated, or ſpotted with different Colours, chiefly Dusky, Black, Brown, Red, and ſometimes Blue.*

CALCEDONY (86). The MOCHO-
STONE (87). OCULUS BELI (88).
ONYX (89). SARDONYX (90). The COM-
MON⁻

(86) Lapis Calcedo-
nius. *This is of the Agat-
Kind; and of a misty Grey,
clouded with Blue, or with
Purple.*

(87) Achates Mocho-
ensis.   Mocho-Stones.
*These are nearly related to
the Agat-Kind, of a clear
horny Grey, with Delinea-
tions representing Mosses,
Shrubs and Branches, in
Black, Brown, or Red, in
the Substance of the Stone.*
Dendrachates, velut Ar-
buscula insignis. *Plin.*
xxxvii. 10.

(88) *The* Oculus Beli
*of the modern Jewellers,
and probably of* Pliny, *is
only an accidental Variety
of the Agat-Kind; having
a grey horny Ground, with
circular Delineations, and
a Spot in the middle of
them, somewhat resembling
the Sight of the Eye;
whence the Stone had its
Name.*

(89) *The* Onyx *is like-
wise an accidental Variety
of the Agat-Kind. 'Tis of a*
dark horny Colour, in which
is a Plate of a blueish
White, and sometimes of
Red; ον́χιον, μικ͑ὴ λευκω̃ ϗ
φαιω̃ παϱαμηλα. Theophr.
*Onyx mixta est ex albo &
fusco parallelis. Laet. Italis*
Nicolo de Quibusdam, A-
chates bicolor. *The said
Colours lying parallel, their
Surfaces terminating, and
meeting in a Plane. The
Lapidaries usually cut this
Stone into two, thro' the
middle of the blueish white
Plate; so that Part of the
White is left adhering to
the darker Colour in each.
When on one or both Sides
the White, there happens
to lie also a Plate of a red-
dish or Flesh-Colour, the
Jewellers call the Stone a*
Sardonyx.

(90) Sardonyx. *The
Lapidaries usually cause
this to be cut, so as to shew
three Colours, Flesh, White
and dark, lying in Planes,
on one another. The* Sar-
donyx *is another Variety
of the* Agat-Kind.

MON-CARNELION (⁹¹). The WHITE-
CARNELION (⁹²). The YELLOW-CAR-
NELION (⁹³). The BERYLL (⁹⁴).

ARTIC. 3. That are in fome
Degree pellucid and tranfparent.

*N. B. The Stones which follow in this
third Article, are thofe which the Lapidaries ufually call* Gemms. *The natural* Conftitution *of thefe having not
been hitherto fufficiently explain'd, I
prefume it will not be thought amifs,
that I premife fomething on this Subject ; fince 'tis from this only, that
their proper Names can be afcertained,
and*

(91) *This has its Name
from its Flefh-Colour* ;
which is, *in fome of thefe
Stones,* paler, *when 'tis
call'd the* female *Carnelion*;
in others deeper, call'd the
Male. *'Tis the* Sardion
Theophrafti, *L.* πεϱι λιϑων,
Sarda Plinii. *L.* 37. *c.* 6.
and the Carneolus *of the
Moderns. The* Italians *give
it the Name of* Cornalina.

(92) *In the* White *of
this, fometimes there is a
very flight caft of Blue.*

(93) *The* yellow Car-
nelion *is very rare.*

(94) *The* Beryll *of our
Lapidaries, is only a fine
fort of Carnelion, of a more
deep bright Red, fometimes
with a caft of Yellow, and
more tranfparent than the
common Carnelion. The*
Beryllus *of the Antients
was a quite different Stone,
of a Blueifh-green Colour :
and probably the fame with
our* Aquamarine.

and their true Ranks aſſign'd. The Baſis, or prime conſtituent *Matter of all of them is, when pure, wholly diaphanous, pellucid, and either* Cryſtal, *or an* Adamantine *Matter, that is more firm and hard. But we find frequently the* Diaphaneity *of this Matter changed and leſſen'd, by Means of a fine metallic Matter, incorporated with the diaphanous, in the original Concretion and Formation of the Stones. By the Acceſs and Mixture of this metallic Matter, I find, by various Experiments and Obſervations, which will appear in their proper Place,* 1ſt. *That the Weight,* or ſpecifick Gravity *of the Stone, is ſomewhat increaſed.* 2. The Hardneſs *of the Stone is varied, chiefly in the Cryſtallin Kind.* 3. The Figure *into which the pellucid Matter naturally ſhoots, is changed, by* Lead *incorporated with that Matter, frequently into a* Cubic Form ; *by* Tin, *into a* quadrilateral Pyramid; *by* Copper, *into very* differing Figures *uncertainly; by* Iron, *chiefly into* Rhomboids. 4. *A Tincture, or* Colour, *is*

impar-

*imparted to the Stone, paler or deeper
in Proportion to the Quantity of the
additional Metal.* 'Tis, *in fome, fo
little, as hardly fenfibly to reflect the
Light, or give any apparent Colour;
when more, it gives a flight pale Colour; when more, ftill a deeper, and
more a faturate : When fo much as perfectly to obftruct all Paffage of the
Light, the Stone quite lofes its Tranf-
parency or Diaphaneity, and becomes*
opake. *Of this we have Inftances in
the* Tin-Pyramids, *the* Iron-Rhombs,
the Lead-Cubes, *and when join'd by*
Copper, *as in the* Lap. Nephriticus, *the*
Malachites, Lap. Lazuli, Heliotropium, Jafper, *and in the* yellow braffy Ludus Paracelfi; *or by* Iron, *as in the
dusky blackifh* Ludus Paracelfi. *When
the metallic Matter is not in fo great
Quantity, as to refufe and bar all Paffage to the Light, but yet fo great as
to reflect it, and fhew a Colour; this,
where* Lead *is the Ingredient, is* Yellow. *Hence the* Topaz, *and the* Jacinth, *which probably, with the Lead,
has an Admixture of Iron, to which*

E   *it*

it owes the mix'd or flame Colour. When Tin *is the Ingredient, the Stone is by it render'd* black ; *as in the* Tin-Grains, *and the* black Agat. *Where* Iron *is the Ingredient, the Stone is by it render'd* red. *Hence the* Carnelion, *the* Beryl, *the* Garnet, *the* Rubin, *the* Carbuncle, *the Amethyſt. Where the Ingredient is* Copper, *if attended with any* Alcali *that may happen to join it, the Stone is* blue; *hence the* Saphire, *and the* Water-Saphire, *if attended with an* Acid, *green ; hence the* Emerald. *When the Ingredient is both* Copper *and* Iron, *the Stone is of a Colour mix'd with Blue and Green. Hence the* Aquemarine ; *when* Copper *and* Lead, *of a Green and Yellow, as in the* Cryſolit.

By the Bounds I am tied up to, I am ſo reſtrain'd, that I can only hint that, from what has been ſaid, may be concluded eaſily enough, that there can be no fix'd and unerring Teſt *or* Standard, *whereby the* Kinds *and* Names *of theſe Bodies may be conſtantly aſcertain'd.*

*tain'd. For, if the metallic Matter
that happen'd to attend the Gemmeous
in its Formation, and to enter the
Compofition of the fame Stone was
various and uncertain, and the Quantity of it as various and uncertain,
there muft, in courfe, be fome Variety
and Uncertainty in the* Colour, *from
which both the* Name *and* Kind *of the
Stone is determin'd: And 'tis from
this that arifes the* Difference *and*
Confufion *that we find among the* Writers of Gemms, *both Antient and
Modern. When the fame Kind of
Stone has its Varieties and Differences,
the Defcribers of it, tho' never fo accurate, muft needs vary and differ;
tho' not fo much as to leave no Rules or
Characters whereby to diftinguifh and
form a Judgment of moft of thefe Bodies. For my own Part, amidft fo
much Darknefs and Confufion, I hope
I have not gone far out of the Way, or
much miftaken my Aim: And what I
fhall offer by and by, relating to Metals, will give fome further Light into
this fo dark and intricate an Affair. I*

E 2 *muft*

muſt not forget to take notice, that even the Placing *and* Diſtribution *of the* metallic Matter *to the ſeveral Parts of the ſame Stone, is not ever uniform, but in one Part a* Red, *or* Iron, *ſhews it ſelf, in another a* Blue, *or* Copper; *nay, in ſome Parts 'tis perfectly* clear *and* tranſparent, *without the leaſt Appearance of* Colour, *or* metallic Admixture. *Of all which* Phænomena, *there are* Inſtances *in my* Collection.

SECT. 1. Thoſe which are tinged with ſome Colour. The To-PAZ ([91]). The HYACINTHUS ([92]), or Jacinth of the Jewellers. The GARNET ([93]). The

([91]) *This is of a yellow or Gold Colour. 'Tis the* Chryſolithus *of the Antients.*

([92]) *This is of a deep rediſh Yellow, approaching a Flame Colour, or the deepeſt* Amber. *The Jewellers have two Sorts, a paler and a deeper, which they call* la Belle, *and which probably may be a Species of the* Carbuncle *of the* Antients. *The Hy-* acinthus *of the Antients was certainly a much different Stone, in Colour Purple, tending to Blue, and ſomewhat reſembling the Flower of the Hyacinth, or* Violet. Conf. Plin. L. 37. *c. 9.* In Amethyſto fulgor violaceus dilutus eſt in Hyacintho. Plin. xxxvii. 9.

([93]) Lapis Granatus. *This ſeems to be a Species of the* Carbuncle *of the* Antients.

Fig. 1. P. 28.

The ROCKY-RUBY (⁹⁴). The BALASS-
RUBY (⁹⁵). The SPINELL-RUBY (⁹⁶).
The CARBUNCLE (⁹⁷). The AME-
THYST (⁹⁸). The SAPPHIRE (⁹⁹). The
WATER-

Antients. The Bohemian *is red, with a flight Caft of a Flame Colour. The* Syrian *is red, with a flight caft of Purple.*

(94) Rubinus rupium. *This is of a Red deep, and the hardeft of all the Kinds.*

(95) Rubinus Balaffi-us. *This is of a Crimfon Colour, with a Caft of Purple, and feems, beft of all the three, to anfwer the Defcription of the* Ru-by *of the Antients.*

(96) Rubinus Spinel-lus. *This is of a bright rofy Red; 'tis fofter than either of the foregoing. Some late Writers fuppofe the Rubies to be defcribed by the Antients, among their* Carbunculi.

(97) *The* Carbuncle *of the modern Jewellers is a Stone of the* Ruby-Kind, *very rare, and of a rich Blood-red Colour. Of the* Aνθραξ, *or Carbuncle of the Antients. See* Theophraft.

& Plin. L. 37. c. 7.

(98) Amethyftus. *This is of a bright Purple.* Ἀμεθυσον οινοπων τιχρόα. Theophr. Uvas maturas Colore refert. *Laet.* Plin. ad vini Colorem accedit in violam definens.

(99) Sapphirus. *The Sapphire is of a bright blue Colour. We have this Stone from the* Eaft In-dies, *where it is call'd* Nilaa *from its Colour;* Nil, *or* Anil, *being the Word they ufe for Indigo, and probably may denote blue in general. It does not appear that this Stone was known to the* Antients. *At leaft there is no Account of it in any of their Books extant. 'Tis certain the* Sapphirus *of Pliny is much different from our* Sapphire; *and his Defcription anfwers to the* Lapis Lazuli. In Sapphiris au-rum Punctis collucet. Plin. *Ita fere &* Theo-phraft. & Ifidor.

WATER-SAPPHIRE (¹⁰⁰). The AQUA-
MARINA (¹⁰¹), of the *Italian* Lapidaries.
The EMERALD (¹⁰²). The CHRYSO-
LITE (¹⁰³).

SECT. 2. Thofe which are
perfectly clear, diaphanous, and without
any Colour at all. CRYSTAL (¹⁰⁴). The
WHITE-

(100) Sapphirus aquea. *This is the accidental* Sapphire, *and is neither of fo bright a* Blue, *nor fo hard as the Oriental.*

(101) *The* Aque Marine *is of a Sea or Blueifh Green. This Stone feems to me to be the* Beryllus *of* Pliny. *That judicious learned Antiquary S. P.* Buonazotti *is of the fame Opinion.* Medaglioni Antichi. *p.* 113. & *alibi paffim.* Pliny *ranks it amongft the green tranflucid Gemms, reprefenting it as related to the* Smaragdus, *but of a Colour lefs brisk, and imitating a pure Sea-Water Green.*

(102) Smaragdus. *This is of a bright Grafs-Green. 'Tis found in Fiffures of Rocks along with Copper Ore.*

(103) *This is the* Topazius *of the* Antients. Vid. Plin. 37. *c.* 8. *'Tis of a dusky Green, with a Caft of Yellow.*

(104) Cryftallus. *This is certainly known and diftinguifh'd by the Degree of its Diaphaneity, and of its Refraction : as alfo of its Hardnefs, which are ever the fame. 'Tis found both lodged in the* S-rata, *and form'd in the Veins, or perpendicular Fiffures of them. In thefe laft, 'tis found ever in Form of an hexangular Column, adhering at one End to the Stone, on the Side of thofe Fiffures, and near the other, leffening gradually, till it termin tes in a Point. This is call'd by the Lapidaries* Sprigg, *or* Rock Cryftal : *And of this Sort* is

is the Iris *of* Pliny, Agricola, *and Dr.* Liſter. *Philoſ. Trans. N°.* 110. *p.* 222. *that fine* Crvſtal *of the* Alps, *as alſo that of* Bohemia, Hungary, *and other Countries, as is likewiſe that found in the Tin-Loads or Veins in* Cornwall ; *tho' a great deal of this is coloured, fouled, and rendred opake, by Admixture of metallic and mineral Matter with the Cryſtallin. Of this Kind of Cryſtal alſo, are the better and larger* Briſtol-Stones, *the* Kerry-Stones *of* Ireland, *the* Pſeudoadamantes *of Authors, and particularly of* A. Boetius de Lap. & Gem. *p.* 120. *The Cryſtal in Form of Nodules, is found lodged ſometimes in the ſtony, but chiefly in the earthy* Strata, *or among the* Gravel, *or other looſe* Rubble *left in a* Train, *by the Water departing at the Concluſion of the Deluge. This Sort, call'd by the Lapidaries,* Pebble-Cryſtal, *is in Shape irregular, and in Form of the common Nodules,* Pebles *and* Flints. *But there is alſo frequently found Cryſtal lodg'd in the* Strata, *in a Form regular, ever*

hexang lar, *which is its diſtinguiſhing and charaċteriſtic Form, and approaching that found in the Fiſſures ; of this Rank are,* 1. Cryſtallus in acumen utrinque deſinens, Cryſtall pointed at both the oppoſite Ends. *Of this I have obſerv'd two Sorts; the one conſiſts of two hexagonal Pyramids, applied Baſis to Baſis.*

Aldrovandus *has an Icon of this Sort, which he calls an* Iris *in his* Muſæum. p. 941. Boetius *has another in his* Hiſt. Lap. & Gem. p. 218. *The other conſiſts of two like Pyramids, but having an hexagonal Column intervening.* Boetius *has there likewiſe an* Icon *of this Sort, as has alſo* Aldrovandus *p.* 989. N°. 2, *where he gives it the Name of* Cryſtallum parvum utrinq; æqualiter mucronatum. *He takes it from* Geſner, De Fig. Lapid. *p.* 19. *who was under ſome Doubt, whether there had not been ſomething of Art uſed in the forming of it ; but that proceeded from his not having made ſufficient Enquiry into theſe, and other not leſs elegant natural Produċti-*

WHITE-SAPPHIRE (^{105}). The DIA-MOND (^{106}).

N. B.

ons *found commonly in the* *Earth. Nor can I quit the* *Subject, without taking no-* *tice, that I have observed* *of both these Sorts, not only* *single and separate,but join-* *ed and united in Clusters,* *several in the same Mass,* *of which, as well as of* *those found single, there* *are various Samples in my* *Cabinets.* 2. Cryftallus Forma globofa folida Py-ramidibus pellucidis per totam fuam fuperficiem exteriorem furrectis ob-fita, the Echinated Cry-ftallin Ball. *I have rarely* *observed any of these* *Balls, that have exceeded* *two inches in diameter.* 3. Cryftallus globofa ex-terne rudis & fcabra, in-tus cava, Cavitatem to-tam habens Pyramidibus Cryftallinis obfitam, the concave Cryftallin Ball. *I have observed of these* *Cryftal Pyramids, tho'* *commonly transparent and* *diaphanous,some that have* *been tinged Yellow, others* *red, others purple. The* *exterior Surface of the*

Crufts and Shells of these *Balls are commonly of a* *brown ruft Colour, confi-* *fting chiefly of a coarfe* *Spar, with some little ear-* *thy, ftony, Mineral, or me-* *tallick Matter incorpora-* *ted with it. I have obferv'd* *these Balls of all fizes,* *from the Bignefs of a Wal-* *nut, to that of the largeft* *Melon. They are seldom* *exactly round, but of a Fi-* *gure nearly approaching it,* *tho' somewhat comprefs'd.* *The three foregoing Kinds* *are found in moft Coun-* *tries ; but I have obferv'd* *them in greateft Plenty* *about* Briftol, *chiefly in* *the Neighbourhood of* Kings-Wefton *in* Glou-cefterfhire.

(105) Sapphirus alba. *The white Cryftalline Sap-* *phire, is so called becaufe* *'tis of full as great specific* *Hardnefs as the Blue, but* *colourlefs, and clear as* *Cryftal.*

(106) Adamas. The Diamond. *This Stone is* *preferable, and vaftly fu-* *perior to all others in Luftre* *and*

N B. *The* Characteriſtic *of the Stones of this Section, I mean, their being* perfectly clear, diaphanous, and without any Colour at all, *does not hold ſo univerſally, but that there are Deviations from it : And they are found ſometimes tinged and coloured. Thus there is Cryſtal, having nearly the ſame Degree of Hardneſs with the common, that is notwithſtanding of a* yellow Hue ; *as likewiſe of a* Red, *of a* Blue, *or of a* Green. *To theſe the* Writers *of* Gemms *have given the Names of* Pſeudo-Topaſius, Pſeudo-Beryllus, Pſeudo-Sapphirus, *and* Pſeudo-Smaragdus, *Conf. A. Boet. de Lap. & Gem. L. 2. c. 72. p.* 219. *Sometimes Part of the Stone is clear, and Part tinged, not only with one ſimple Colour, but perhaps with two, or more, all different. In the ſame Manner, the oriental* Sapphire, Topaz, Amethyſt, Emerald, and Ruby, *are all of the ſame Hardneſs.*

F *There*

---

and Beauty: *As alſo in* Hardneſs, *which renders it more durable and laſting, and therefore much more* valuable *than any other* Stone. *Such it has been reputed in all Ages, and by all Nations : And indeed the very Top-Jewel of the whole Creation.*

There are Diamonds *tinged with* Yellow: *Others with* Red, Blue, *or* Green, *tho' these last be very rare. The Tinctures and* Colours *of these, as of all other Gemms, and Stones, are owing to the Principles assign'd above; I mean metallic and mineral Matter, incorporated with the diaphanous, at the first Formation of the Body. That they actually are so, and the Thing really Fact, I have given several Instances in the* Catalogues and Accounts of the Fossils, both of my *English,* and Foreign Collections; *as also various Proofs from* Trials *in the* Fire, *and Illustrations by* Chymical Experiments *in my* Art of Essaying, *and some other* Papers.

Class 3. SALTS.

OR Bodies friable and brittle, in some degree pellucid, sharp, or pungent to the Taste, dissoluble in Water, but, after that is evaporated, incorporating again, crystallizing, and forming themselves into angular Figures.

The

The Fossil-Salt (¹). Sal-Ammo-
niac (²). The Tincal (³) of the *Per-*
F 2   *fians.*

(1) Foſſil, *or* Rock-
Salt, *and* Sal Gemme-
um ; *ſo call'd from its
breaking frequently into
Gemm-like Squares. Theſe
two Salts differ not in Nature or Property from each
other :* Nor *indeed from
the Common-Salt, of the
Salt Springs, or from that
of the Sea, when all are
equally pure and free from
extraneous Matter.*

(2) Sal Cyrenaicum ſeu
Ammoniacum nativum
veterum. Plin. *L.* 37. *c.*
7. *&* Dioſcorid. *L.* 5. *c.*
126. *according to* Fr. Im
perati, De Foſſil. *p.* 20.
'*Tis found ſtill in* Ammo-
nia, *the Country mention'd
by the Antients, and from
which it had its Name.
His Account is confirm'd
by Mr.* Jezreel Jones, *who,
having liv'd ſometime in
the Kingdom of* Morocco,
*and made himſelf Maſter
of the Language, was, at*

the Expence of Dr. Teni-
ſon, *late Lord Arch-Biſhop
of* Canterbury, *my Lord*
Somers, *Sir* Hans Sloane,
*my ſelf, and ſome o-
thers, Lovers of natural
Hiſtory, ſent, about the
Year* 1705, *into the Country thereabouts, to make
Obſervations and Collections : And he found this
Salt, native, in the Earth,
in ſeveral Places. This,
as likewiſe* Tincal *and*
Natron, *are not ſimple Bodies ; but different Salts
concreted with a ſmall Admixture of ſome terreſtrial
Subſtance.*

(3) *The* Tincal *of the*
Perſians. *This ſeems to
be the* Cryſocolla *of the*
Antients ; *and is what our*
Borax *is made of. The*
Indians *of* Bengal, *where
there are great Quantities
of it brought d̄wn the*
Ganges, *call it* Swagar.

fians. NATRON (4). The NITRE (5)
of the Moderns, or Petre-Salt. The
FOSSIL-

(4) *This is the* Νίζρον, Nitrum *of the* Egyptians, *and had its Name from* Nitria *a Province of that Country in which chiefly 'twas found; but 'tis call'd there at prefent* Natron, *or* Latron. *Dr.* Hunting-ton, Epift. *p.* 68. " La-" tron Aquis in Nitria " Ægypti deferto, —Su-" pernatat ad modum Gla-" ciei, cui maxime fimile " eft, fed durius, rubef-" cens. Carnem infulfam " gratam reddit. *p.* 69. " —Defertum, quod o " lim Nitriæ, hodie S. " Macarii dicitur, Locus " eft fteriliffimus—. A-" qua falfa eft. Arbores " nullæ funt, neque Ar-" bufta, nullæque præter " Alcali, Herbæ. *Conf.* " *Difc. of Vegetation.* " *Philof. Tranf. June.* " 1699. Tenet equidem " Salis lacum æque ac " Nitri, nec non Lapi-" dum, Calcis, & Margæ " Fodinas. *There have been made feveral Experi-ments upon* Natron, *by the*

Operator of the Acad. des Sciences, of which there is an Account in Dr. Tournfort's Preface to his Hift. des Plantes aux En-virons de Paris. p. 11. Conf. 37.
(5) Nitre, while in its native State, is call'd Pe-tre-Salt ; when refin'd, Salt-Petre. 'Tis of Ufe in Vegetation. Vid. Difc. of Vegetation. Ib. and that it might be every where ready, and at hand, to ferve that important End, 'tis fcattered about, and mix'd with the Earth, near the Surface, on which Ve-getables are produced in all Countries quite round the Globe. But 'tis found like-wife lying very fhallow, and but juft underneath the Turf, in much the greateft Quantity that we know of, about Patnafs, in the nor-thern Parts of the Kingdom of Bengal ; whence we have ours. Father Fœlix White, was, on Account of his Miffion, fome Time in the Country where this Salt is
got

FOSSIL-ACID-SALT (⁶), ſeldom found ſimple and pure, but in Form of SUL-PHUR, ALUM, or VITRIOL.

Claſs 4. BITUMENS.

OR Bodies that readily take Fire, and yield an Oyl; and that are ſoluble in Water

CA-

---

got; *and he favour'd me with a Relation of the incredible great Plenty of it there, the Manner in which it lies, and all Circumſtances of it*; *but that Relation is too long to be inſerted here.*

(6) **Sal** Acidum Foſ-file. *This is indeed the Baſis of* Sulphur, Alum, *and* Vitriol. *The ſimple Salt, extracted out of any of the three indifferently, is the ſame*; *and is capable of conſtituting either of the* other; *with the Addition of a ſmall Proportion of a* bituminous, cretacious, *or* metallic *Matter.* Sulphur *is produced by incorporating an* oily *or* bituminous *Matter with this* Salt. Alum *is produced, by joining a* cretaceous *or other like earthy Matter with it.* Vitriol, *by Addition of a* metallic *Matter. If* Iron *be made uſe of, the* Vitriol *will be* green; *if* Copper, Blue.

CAPVT 1. Thofe that are liquid.
NAPHTHA (¹). PETROLEUM (²). BAR-
BADOES-TARR (³).

CA-

(1) Naphtha, νάφθα, Di-
ofcor. *L* 1. *c.* 101. Strabo, Geogr. *L.* 16. *repre-*
fents it as a Liquation
of Bitumen. *It fwims on*
the Top of the Water of
Wells and Springs. Salmaf.
Exerc. in Solin. *That*
found about Babylon is in
fome Springs whitifh, tho'
it be generally black, Strabo, Ib. *and differs little*
from Petroleum.

(2) Petroleum *is a li·*
quid Bitumen, *Plin.* xxxv.
15. *black, floating on the*
Water of Springs. Such
is that of a Spring rifing
at the Foot of a Mountain

near the Sea, in the Ifland
Zant, *mention'd by the*
Antients. Sir Geo. Whee-
ler *has alfo given an·Ac-*
count *of it in his* Voyages,
p. 48.

(3) Oleum Terræ Bar-
badenfe. *See* Ligon's *Hift.*
of Barbadoes. *It differs*
little from the Petroleum,
found floating on a fmall
Spring at Pichford *in*
Shropfhire, Camden. *and*
in other Springs of England, *and of* Scotland. *Sir*
Robert Sibald Prodr. Nat.
Hift. Scotiæ. *Part* 2. *L.*
4. *c* 4.

CAPVT 2. Thoſe that are call'd Bɪ-
ᴛᴜᴍᴇɴ (4). Pɪssᴀsᴘʜᴀʟᴛᴏɴ (5). Aᴍ-
ber (6). Iᴇᴀᴛ (7). Cᴀɴɴᴇʟ-Cᴏᴀʟ (8)
Pɪᴛ-Cᴏᴀʟ (9), *Stone-Coal, Quarry-Coal,*
Sea-Coal.

Claſs

(4) Bitumen. Ἄσφαλλος. Dioſcorides *L.* 2. *c.* 99. *mentions it as found about* Sidon in Phœnicia, *in* Zant *and* Sicily, *but pre-fers that of* Judæa *to all* others. Dioſcor. Strabo, *and others of the Antients, aſſert, that both* Bitumen *and* Petroleum *are found plentifully about* Babylon ; *which very remarkably con-firms the* Moſaic *Account of the Uſe of it as* Mortar, *in building the Tower of* Babel, Gen. xi. 3. *Nay, the Buildings of old* Babylon *were, like that Tower, of* Brick *cemented with* Bitu-men. Strabo, *L* 16. Plin. *L.* 35. *c* 15.

(5) Πισάσφαλος *was found in the* Ceraunian *Moun-tains of* Apollonia, Dioſ-cor. *L.* 1. *c.* 100. *The an-*

tient Greeks *gave the Name of* Πισάσφαλλος *to* the liquid, *as well as to* the ſolid Bitumen.

(6) Succinum Lyncu-rion *Demonſtratus ap.* Plin. *L.* 372. *Græcis,* ἤλεκΊρον : *Germanis Vete-ribus,* Gleſum. *Tacit. de Morib. Germ. c* 45. *Ara-bibus,* Karabe.

(7) Gagates. ΓάγαΊος. Dioſcor. v. 146. Gagates Lap. niger eſt, planus, pumicoſus, non multum a Ligno differens, levis, fragilis. Plin. *L.* 36. 19.

(8) *This ſeems to be the* Lapis Ampelites *of the Antients.* Bitumini ſimil-lima eſt Ampelites. Plin. XXXV. 16. Ἀμπελῖτις Dioſ-cor. *L.* 5. *c.* 181.

(9) Carbo foſſilis Car-bo ſaxeus. λιθάιθραξ.

Class 5. MINERALS.

OR Bodies nearly related to Metals ; as having some Properties in common with them, being particularly ponderous, and splendent with a metallic Brightness.

CAPUT 1. Those that are fluid. NATIVE-MERCURY, or VIRGIN-QUICK-SILVER (¹).

C A-

---

(1) Dioscorides *takes notice of* Quick-Silver *that was native, and found in the Earth fluid, free, and without Mixture : and calls it* ὑδράργυρος καθ ἑαυτὸν, Mercury *is a Mineral of very singular and peculiar Nature, and differs from all others in keeping constantly a fluid Form, when pure, separate, and unmix'd. Nor can it ever be fix'd, or brought to Consistence and Solidity, by any Art whatever. It amalgams with all Metals, except only Iron, and is susceptible of a more consistent Form, when united with Nitre, Alum, or o-ther acid Salts, and with* Arsenic, *or Sulphur. But, when disengaged from them, and separated again, it ever appears in its original natural Condition, and fluid as before. Would our* Alchymists, *who work much on* Mercury, *reflect rightly on this, 'twould put an End to their troublesome, expensive, and delusive Amusements. 'Tis call'd* Χυτὸν ἄργυρον *by* Theophrast. de Lap. ὑδράργυρος *by* Dioscor. L. 5. c. 110. Hydrargyrum *by* Pliny, L. 33 c 8. Argentum Vivum, *ibid.* L. 33. c. 6.

CAPVT. 2. Thoſe that are ſolid, and will melt in the Fire, but are not du£til or malleable. NATIVE-CINNA-BAR(²). NATIVE-YELLOW-ARSE-NICK (³). NATIVE-RED-ARSENICK (⁴):

G   The

(²) Cinnabar *is the Ore out of which* Quick-Sil-ver *is drawn, and conſiſts partly of a mercurial, and partly of a ſulphureo ochre-ous Matter.* Dioſcorides, *L.* 5, *c.* 109, 110. *calls it* "Αμμιον, *or as other Copies have it,* Μίνιον, *and makes a Diſtinction betwixt this, and* Cinnabar, Κιναβαει. *The former, he ſays, they had from* Spain, *the latter from* Africa; *and probably there might be ſome Diffe-rence betwixt them; but, by the Properties and Uſes he aſcribes to each, they ſeem to be of the ſame Kind. At leaſt* Pliny *tells us ex-preſly ſome of the* Greek *Writers called that* Cinna-bari, *which the* Romans *called* Minium, *and out of which they extracted their* Hydrargyrum. *Others cal-led it* Miltos; " Milton " vocant Græci Minium,

" quidam Cinnabari L. 37. c. 7. conf. c. 8 I. Ant. Saracenus *Not. in* Dioſ-corid. *corrects the Place, and ſubſtitutes* Ammion, *but without Reaſon,* Μιλος *being the Word uſed by ſome of the* Greek *Wi riters, and particularly by* Strabo, *conſtantly.* Theophr. *L. de Lap. uſes only the Word* Κιναβαει; *ſo that 'tis plain, that ours and the Antient* Cinnabar *is the ſame*

(3) Arſenicum Aure-um nativum. 'Αρσίνιχον *Di-oſcorid. L.* 5. *c.* 121. Auri-pigmentum, *Plin. L.* 35. *c.* 6. *& L.* 33. *c.* 4. Ar-ſenicum, *L.* 34. *c.* 18.

(4) Arſenicum rubrum nativum, Σαρδαραχη Dioſ-corid. *L.* 5. *c.* 122. San-daracha. Plin. *L.* 34. *c.* 18. *This is mention'd by* Agricola de Nat. *Foſſil. L.* 3. Fr Imperati. *de Foſſ. p.* 29. Ol. Worm. Muſ. *L.* 1.

42   *A Method of Fossils.*

The PYRITES (5). The MARCASITE (6).
COBALT

L. 1. Sect. 1. c. 12. and others. The Hungarian Sandaracha *is of an Orange Colour : But that from* East India *of a deeper Red. I have Samples of each ; but both are very rare.*

(5) Pyrites. *This Body ever contains more or less of the* Sal acidum, *that is incorporated with an oleose or bituminous Matter, and so constitutes a* Sulphur. *This renders it so apt to give* Fire, *from which it has its Name* Πυρίτης. πῦρ. Fire. *It sometimes contains a cretaceous, or ochreous, and constantly a metallic Matter, in it : In proportion, as any of these prevail in Quantity, and come forth incorporated with the Salt, it appears in form of* Sulphur, Alum, *or* Vitriol. Conf. Not. ad Class. 3. *supra. I never met with any* Pyritæ *that held* Lead *or* Tin. Copper *there is in some of them ; and* Iron *in all : but the Quantity of it is not considerable. In those that hold most of*

it, when the Salt is drawn off, the Iron *usually constitutes about $\frac{1}{8}$ of what remains. They all hold an extremely small and inconsiderable Quantity of* Gold, *and some few of* Silver.

(6) Marcasita. *The Writers of Minerals generally give the Name* Pyrites *and* Marcasita, *indifferently, to the same sort of Body : And indeed they both agree in some Things. But I choose rather to restrain the Name of* Pyrites *wholly to the* Nodules, *or those that are found lodged in* Strata, *that are separate, and not a Part of, or depending on the common Matter of the* Stratum. *The* Marcasite, *on the contrary, is Part of the Matter that either constitutes the* Stratum, *or is lodged in the perpendicular* Fissures *of the* Strata. *The* Marcasite *frequently holds* Arsenic ; *which the* Pyrites *does rarely, if ever. There is* Sulphur *in all* Marcasites : *And* Antimony *and* Bismuth *in some. The* Metalls *they yield*

Cobalt (7). Calamin (8). Antimony (9). Tin-Glass (10). Zink (11). Wad, or Black-Lead (12).

Clafs 6. METALLS.

O r Bodies that are ponderous, fplendent, folid, will melt in the Fire, and are ductil or malleable.

G 2   1. Gold.

yield are chiefly Copper, Iron, *and* Tin. *When any of thofe Metalls were in confiderable Quantity, thefe Bodies lofe the Name of* Marcafites, *and are call'd* Ores. *In* Cornwall, *and the* Weft, *they call them* Mundick, *in which there is commonly* Copper, *or* Tin, *and fometimes* Iron. *But* Mundick *abounds fo much in* Sulphur, *that the Metalls are very difficult to be parted. Being run down all together, they compofe a Kind of* Bell-Metall, *ufed by fome for making Bells, Mortars, and the like.*

(7) Cobaltum, *a* Marcafite *frequent in Saxony*.

It is plentifully impregnated with Arfenic, *contains* Copper, *and fome* Silver. G. Agricola, In Bermanno *p.* 690. 701. Ol. Wormius, Mufæum. *p.* 128. *and the reft of the Writers of Minerals take this for the* Cadmia *of the Antients. Being fublim'd, the Flores are of a blue Colour. This the* German *Mineralifts call* Zaffir.

(8) Lapis Calaminaris.

(9) Antimonium S. Stibium.

(10) Bifmuthum.

(11) Speltrum.

(12) Nigrica fabrilis, Merreti. Pinax Rer. Nat. Britan.

1. GOLD. 2. SILVER. 3. COPPER. 4. IRON. 5. TIN. 6. LEAD.

FROM what I have deliver'd on another Occaſion † concerning the Confuſion that Things lye in under Ground, and the various Combinations of Metalls amongſt themſelves, and their Mixture with almoſt all other Sorts of terreſtrial Matter whatever, may readily be concluded how difficult a Task it is to deſcribe the Ores of them, and diſtinguiſh each from other. I have for ſome Years been carefully examining thoſe found in *England*, and procured Samples from moſt other Parts of the known World. What Rules and Diſtinctions of the various Sorts I have been able to make, I ſhall next deliver as clearly as the Bounds I am tied to will permit.

1. GOLD, *Aurum*, χρυσὸς. This Metall conſiſts of Parts ſo infinitely ſubtil and fine, that when 'twas all in ſolution, and

† *Nat. Hiſt. Earth. Pt.* 4.

thoſe

thofe Parts divided, and abfolutely feparated each from other, which was the
Cafe at the Deluge, they would be fo eafily agitated and difperfed about every
where, that 'tis not ftrange that we find
more or lefs of this Metall incorporated
with almoft all Kinds of terreftrial Bodies whatever. But, as it feems, the main
Bulk of it, before the Diffolution at the
Deluge, lay chiefly in fome particular
Places, it fubfided again in them; and
there chiefly it muft of courfe be at this
Day found. 'Tis interfpers'd, mix'd, and
incorporated with the *Strata* of the
Earth or ftony Matter; and the Particles
of it commonly fo fmall, as not to be dif-
cernible; but fometimes they lye fo clofe
and thick, as well to compenfate the Labour and Expenfe of wafhing away the
Earth wherein they were lodged; and the
ftony Matter, after 'tis beat, broken, and
finely reduc'd: For when this is feparated by Means of Water, and decanted off,
the Gold, being ponderous, all readily
fubfides to the Bottom; by which Means
'tis collected and preferv'd. In this Manner 'tis wrought in the Mines of *Cânia*,
and

I

and other Parts of *America* ; in *Achin*, and other Parts of *India*, and the Eaſt ; and in the Mid-land Parts of *Africa*. Gold is found likewiſe in the *Strata* in bigger Particles, Maſſes, and Lumps of various Sizes. The largeſt that I have ſeen of Gold thus ſeparate and pure, taken out of a *Stratum*, weigh'd near three Ounces. But ſuch are very ſeldom met with ; though there are Accounts of Princes, and great Perſons, living in the Countries where the Gold is got, that have much larger Lumps and Nodules of it. Beſides, the Gold thus found in the *Strata*, 'tis likewiſe met with in the Veins and perpendicular *Fiſſures* of them, either incorporated with the Sparry, Mineral, or metallic Matter repoſited there, or ſeparate and pure. This laſt is ordinarily found adhering to the *Spar*, and run into Form of *Threads* and *Grains* ; whence it has obtain'd the Name of *Aurum nativum fibroſum, & granulatum.* Sometimes ſuch is found concreted and affix'd to the Stones on the Sides of the *Fiſſures*. Of all theſe there are Samples in my Collection.

THEN

THEN there is found Gold in Form of
Duft, Powder, Grains, and Lumps, at,
or near the Surface of the Earth; but
chiefly on the Shores and Strands of the
Rivers, and on the Sides, and at the Feet
of Mountains. This is all wafh'd forth
of the Earth by the natural Action of
Water; that found about Rivers,
partly by their common flowing and
wearing of the Banks, and partly by their
more forcible Action, when there are
great Tides, and Inundations; the Water wafhing away the lighter terreftrial Matter, and fo bareing, uncover-
ing, feparating, and leaving behind the
more heavy Metallic. In this Manner
Gold has been found in all Ages; not
only in the Countries where it abounds,
and there are Mines of it, but in *Greece,
Spain, Hungary,* and other Parts where
there are none. That found about Mountains is wafhed forth by the Falls of
Rains. Thefe in fome Countries are
very great, powerful, and fall conftantly
at certain Seafons. They wafh away the
earthy, and even the loofer ftony Matter;
by

by which Means they difclofe the Gold:
And where it happens to be repofited in
any confiderable Quantity within, after
the Rains are over, 'tis found left on the
fides of the Mountains, and about the
Bottoms of them, in Plenty proportio-
nate to the Greatnefs and Duration of the
Rains. This is the Cafe of the Gold of
Quito in *Peru*, and of moft of that we
receive from *Guinea*, and other Parts of
Africa, where the Mountains, chiefly
thofe up in the Country, abound with
this Metall. Upon Trials in the Fire, I
have found fome of this *African* Gold fo
rich and pure, as to yield 97 or 98 *per
Cent.*

2. Silver. *Argentum*, Ἄργυρος. This
Metall is found in the Veins and Fiffures
of the *Strata*, fometimes native and
pure, adhering either to the Stone on the
Sides, or to the fparry, or other mineral
Matter in the Veins, in various Forms,
e. gr. of *Hairs* or *Threads*, finer or thicker,
of branch'd *Shrubs*, and of *Feathers*; as
alfo fometimes of *Grumuli*, *Maffes* and
Lumps ; from which Forms it has obtain'd

obtain'd the Names of *Argentum Capil-
lare, Fibroſum, Arboreſcens, Plumoſum,
Grumulatum, Concretum.* The fineſt Silver Ore of *Saxony,* is incorporated with
Sulphur and *Arſenick,* which together
impart to it a ruddy Hue. This Sort the
Miners there call *Rothgultig-Ertz.* But
in *Germany, Hungary, England,* and
o:her Parts of *Europe,* the Silver is ſepa-
rated from the Ore of *Lead,* chiefly, that
ſhining, ſparkling Sort, that the Miners
call *Steel-grain'd Ore.* I have, upon
Trial, extracted from ſome of this, one
15th Part of Silver; but ſo great a Proportion is not common.

3. *Copper, ·Æs,* κύπριον, * *Cuprum,*
κάλχος. The principal Varieties, and
Sorts of the Ores of this Metall, are the
Pale-grey, the *Black,* the *Red,* the
Gloſſy-Purple, the *Blue,* the *Æruginous*
or *Green :* The better Sort of *Mundick,*
or the *Marcaſitic Yellow,* ſhining, Braſs-
like Copper-Ore; the *fibrous,* or ſtriated,
and the *ſparkling or Steel-grain'd.* Be-
        H   ſides,

---

*Quoniam in Inſulâ Cypro *copioſè prognatum.*
Vide Plin.

fides, this Ore is fometimes found native
and pure, in Form of *Threads ;* of *Shrubs*
in *Flakes* and *Plates,* fome folid and
continuous, others porous ; in *Grains,*
Maffes, and *Lumps.* Thefe pafs all in
general, by the Name of *Virgin-Copper-*
Ore : And many of them are fo pure, as
to be flexile and malleable, like the refin'd Metall it felf. *Terre-verte, Terre-*
bleue and *Ultramarine,* which is the
blue Part of the *Lapis Lazuli,* all contain fome Copper in them. The *Lapis*
Armenius, † is really a Copper Ore, but
generally very poor ; tho' there is, in my
Collection, fome fo rich, that it yields
one third Copper.

4. *Iron, Ferrum,* Σίδηρος. I have obferved above, that *Gold* is found intermin-
gled with the fandy, earthy, or other common Matter of the *Strata.* I fhould have
taken notice above, that *Copper* is found
fo too ; and renders the Stone wherein it
is contain'd, of a Green or a Blue, or a
ruddy coppery Hue. Iron is frequently
found

---

† Diofcorid. περι υλης ιατρ. *L.* v. *c.* 105.

found in the ſame Manner in the *Strata;*
and, when in Quantity, imparts a ruddy
or ferruginous Colour to them : But neither Silver, Tin, nor Lead, are ever
found in any conſiderable Quantity in the
Strata. The harder red ochreous Iron-
Ores, paſs by the Name of *Rudle;* the
ſofter by the Name of *Smitt.* There is
more or leſs of this Metall likewiſe incorporated with the ferruginous cruſtated
Bodies, the ochreous *Ruſt-coloured-Eagle-
Stone,* the *Bezoar Mineral,* the ferruginous *Geodes,* and the *Enhydros.* There
is found Iron-Ore, in Form of *Ludus
Helmontii,* particularly in *Monmouth-
ſhire,* where this Sort is call'd *Pin-Ore.*
The reſt of the Sorts are, The *ſmooth-
grain'd Iron-Ore,* which ſtrikes Fire,
and breaks much like a Flint, but is of a
ruddy Colour: The *Hæmatites,* or *Schi-
ſtos,* † which is of a ſtriated, or fibrous
Texture, and the *Iron Stalactitæ;* ſeveral of theſe naturally united into one
Sheaf, paſs by the Name of *Bruſh-Ore.*
The *Rhomboid-Iron-Grains.* I have ſeen,

H 2 in

---

† Dioſcorid. *L. c.*

in the Mines of the Forreſt of *Dean*
ſome little Iron Ore, in the Veins,
ſhut into a *Ramoſe,* or *arboreſcent Form.*
Iron is ſeldom found native and pure. I
never ſaw but one ſmall ſample of it,
which came from *Saxony.* But ſome of
the richeſt Ores of this Metal, both the
Engliſh, and thoſe from *Germany,* being
reduc'd to a very fine Powder, the purer
Iron Grains follow and obey the Loadſtone. *Magnes* the *Loadſtone* alſo holds
a little Iron, and is ſometimes found in
the Veins, along with the Ores of that
Metall ; as is alſo the *Magneſia,* or *Manganeſe:* And indeed this differs little
from the *Hæmatites,* only that it is
poorer, and yields leſs Iron. *Smiris,* or
Emery, has likewiſe uſually in it ſome
ſmall Admixture of *Iron.*

5. *Tin, Stannum,* κασσίτεϱς. There
is of the Ore of this Metall got in leſſer
Quantities, in *Saxony,* and in *Bohemia,*
and ſome on the Coaſts of *Malabar* in the
Eaſt Indies. But no Part of the World
yields ſo much of it as *Cornwal,* nor ſo
rich and good. This is the only Product
                                        of

of the Nation, that was fent Abroad, before the *Romans* came hither. The *Britains* had, from the remoteft Antiquity, carried on a Trade with the *Phænicians* in this Commodity. They fent it in Boats, the beft they had in thofe early barbarous Times, made of Wicker, and cover'd with Hides of Beafts, to the Ifle of *Wight,* and thence, to the oppofite Coafts of *France,* whence 'twas carried over Land to *Marfeilles*; where the *Phænicians* bought it, and tranfported it to all Places with which they had Commerce. The principal Sorts of *Tin-Ore* are the *Pale,* near *White,* the *Grey,* the *Brown,* the *Ruddy*; but the beft and richeft is the *Black.* I have never feen, nor heard of any *native pure Virgin Tin.* The *Tin Grains,* or *Tin Corns,* as the Miners call them, are the richeft, and yield about half Metall. There are fometimes a very few Sparks of Metall in that fort of Stone that the Tinners call *Pedancarn,* and in that which they call *Growan.* This laft is a gritty Stone, of various Colours, and of *Talky* Conftitution, having *Micæ* in it. The *Tin-Veins,* or as
the

the Miners call them, *Loads,* are either
in *Strata* of *Growan,* or of that grey,
Talky, Slaty Stone, that the Tinners call
Killas, Raze, or *Delvin.* The greateſt
Quantity of Tin-Ore is found in the
Loads ; but there is of the very ſame
Sorts, found likewiſe in the *Shoads* or
Stream-Works. Theſe are Trains of Ore,
Spar, and other Minerals, that were
waſhed down from the *Loads,* by the
Water departing at the End of the *De-
luge.* Mr. *Carew,* in his *Survey of Corn-
wall,* has given ſome Account of thoſe
Shoads ; but I have obtain'd a much ful-
ler, more particular, and ſatisfying Ac-
count from ſome of the Gentlemen of
that County, and Stewards of the Tin
Mines, that have been curious, and taken
Pains in making accurate Obſervations on
the State of Things there.

6. *Lead, plumbum,* μόλυβδος. The various Names and Diſtinctions of this Ore,
uſed by the Workmen, are, the *Potters*
or *Blue,* the *Grey,* the *greeniſh Yellow,*
the *Talky,* the *Stony,* the *Cavernous,* the
porous Sort, call'd on *Mendip, Honey-
Comb*

Comb Lead-Ore, the *Star-grain'd* Lead-Ore, the *ſtriated,* or *Antimoniated* Lead-Ore, the *ſparkling* or *Steel-grain'd;* this commonly yields more or leſs Silver, and is what *Dioſcorides,* and the Naturaliſts after him, call *Molybdæna: Pliny, Galena.* The *White* ſemi-diaphanous Lead-Ore, generally fibrous, but ſometimes *flaky* or *plated.* The *Ericoid-*Lead-Ore, found concreted into the Form of the *Ramoſe* Moſs, or, as ſome fancy, of *Heath* or *Erica,* whence it had its Name. The *Diced* or *Cubic* Lead-Ore. The *Saxon* Mineraliſts ſometimes find Lead in the Veins, that is *native* and *pure :* But I never ſaw any except one Sample that was fetch'd for me, with ſeveral other Ores, from *Mendip,* by Mr. *John Hutchinſon,* a Man brought up from his Youth in Mines, in the Service of Dr. *Bathurſt* and Mr. *Squire.* Mr. *Auditor Harley* and I borrowed him of his Grace the *Duke of Somerſet,* whoſe hired Servant he then was, and ſent him into the Weſt, to make Searches and Collections for us.

I can-

I cannot well difmifs the Subject, without taking fome brief Notice, of thofe which the Miners call *Mock-Ores*, or *Samples of Veins*, as judging them to be *Signs* of *Ores* lodged fomewhere near. That does not always happen, tho' indeed they are commonly found at the Tops of the metallic Veins. The greateft Part of thefe are very light, porous, and friable; but fome there are that are folid, and fo ponderous, that they certainly hold Metall, tho' fo intimately incorporated with the Mineral Ingredients of the Mafs, as not to be extricated, or feparated from them, by any Procefs yet found out. I fhall conclude, after I have given the Names of the principal Kinds. Thefe are *Mock-Lead, Blind, Blend, Black-Talk,* or as the *Germans* call it, *Sterile-Nigrum. Mock-Tin,* or *Cockle. Mock-Copper,* or *Goffens,* a Cornifh Mineral, as is alfo *Mundick,* a fort of braffy Marcafit there. *Mock-Iron,* or *Call,* likewife the Product of *Cornwall. Mock-Hæmatites,* Mock-Sparry, and Talky-Ores.

F I N I S.

LETTERS

Relating to the Method of Foffils.

LETTER I.

TO
Sir Isaac Newton.

A Letter fent along with the Method of Foffils, giving an Account of the Things needful and preparative to the drawing up fuch a Method. The Difficulties of it, and its Ufes.

Sir,

 SEND you, with this Letter, a *Tract* relating to the Method of *Foffils*; which, if not your own, is wholly owing to you; it being begun, carried on, and finifhed at your Requeft. It is indeed a Work,

I   tho'

tho' fmall in Bulk, I hope, not altogether without its Ules. For as it may be of Service, at leaft to thofe who have now, for fome Time paft, taken Pains in obferving and collecting Foffils, fo it may contribute fomething towards the Advancement of the Science it felf. For a right methodizing of natural Things, and a Diftribution of each into their Claffes, according to their natural Properties, and mutual Agreement amongft themfelves, conduces very much to the more eafy and certain Knowledge of them. For which Reafon, feveral very learned Men of late Years, have happily imployed themfelves, and fpent much Time and Labour, in reducing all Kinds of Animals and Vegetables into Method. But Foffils, of however great Worth and Importance, have been much neglected, and left wholly to the Care and Treatment of Miners and meer Mechanicks. 'Tis on this Account that thefe, having not been yet fufficiently made known and diftinguifh'd, have lain hitherto in the Dark; till being, Sir, at your Command,
                                                brought

brought forth to Light, I now diſplay, and lay all open to your View.

The Reaſon that there has been a ſo much greater Progreſs made in di-geſting and methodizing Animals and Vegetables, is, that they are more fre-quently in View, better, and more readily known. For, in thoſe Bodies, the Marks and Characters, by which the principal Kinds, and ſubordinate Species are di-ſtinguiſhed, being ſo manifeſt and ap-parent, their Affinities or Differences may be diſcerned with Eaſe, and almoſt at firſt Sight. Whereas, Minerals are of a deeper, and much more abſtruſe and dif-ficult Inquiry. Of this I ſhall produce one or two Inſtances. As the exterior na-tive Complexion, in Samples of even the ſame Kind of Mineral, is commonly very different; ſo likewiſe muſt the interior Conſtitution be, by reaſon of the various extraneous Matter that is commonly in-corporated with it in its firſt Concretion. Nor is there a leſs Diverſity in the Site of Minerals, their Place, and in the Variety

of Matter, among which they are found
lodged and repofited in the Earth.

THAT I might therefore the better ex-
tricate my felf from thefe fo great Perplexities, and come to fome Certainty in
this Affair, I propofed feveral Ways of
Examination and Trial, in order to difcover the Nature of fuch Parts in thefe
Bodies, as do not immediately fall under
the Senfes. The firft of thefe was, to
find out and afcertain the various Degrees
of the *Hardnefs* of each. The next,
to make accurate Obfervations of their
various *fpecific Gravity.* Finally, I tried
each by *Fire,* and a *Chymical Analyfis,*
in order to difcover whether they would
emit an *Halitus* or *Vapour,* or a *Smoke,*
or a *Flame:* Whether they would yield
an *Oil,* or a *Salt :* Whether they would
be reduced to a *Cinder,* or a *Calx :*
Laftly, whether they would run into a
Vitrum, or into fuch a Mafs, as the Me-
tallifts are wont to call a *Regulus.* Befides, as I am not forward to rely on my
own Abilities, well knowing how little
they are, I thought it proper, in fo ob-
                                    fcure

fcure and intricate a Subject, to confer
with fome others, who were well vers'd
in the Knowledge of Minerals, particularly Mr. *Stoneftreet*, whofe Sagacity in
fearching into natural Things, and Succefs in methodizing them, I had been
long acquainted with. Neither would I,
after all, have thus offer'd thefe my At-
tempts to a Perfon of your Judgment,
without having firft had the Approbation
of thofe others, who are moft defervedly
in Efteem for their Knowledge in thefe
Studies. If I find what I have here laid
before you be not unacceptable, as it will
be the higheft Satisfaction to me, fo will
it encourage me, if ever I am fo fortu-
nate as to have leifure to lay before you,
and, if it be fo happy as to have your
Approbation, to publifh a Natural Hiftory of all the Sorts of Foffils, founded on
Reflections made upon thofe I have collected, and the Obfervations that I have
made on others from abroad.

I am,   *c.*

LETTER II.

To Sir J o h n H o s k y n s Baronet.

The Study of Foſſils never hitherto redu-
ced to Rule, nor any Form of Art.
The Writers, both the Antients, and
thoſe of later Times, have confounded
Things buryed in the Earth, with the
natural conſtituent Parts and Produ-
ctions of it. Theſe diſtinguiſh'd, the
Ranks of each adjuſted, and Foſſils *di-*
vided into Extraneous *and* Native.

S i r,

I Have little to value my ſelf upon,
beſides the Goodneſs I am perpetually
receiving from my Friends, and the fa-
vourable Opinion they are pleas'd to en-
tertain of my Studies. Nor does any
Thing in Life afford me ſo ſenſible a Plea-
ſure, as the Reflection that I am able to
do any Thing that is not thought wholly
unworthy of Acceptance with Men of
the

the Character of thofe you mention. 'Tis
particu arly no fmall Satisfaction to me, to
be fo far honoured with the Friendfhip of
Mr. *Aglionby:* And, that a Man of his
Goodnefs, and extenfive Knowledge, is
pleas'd to think me capable of inlarging,
or making any Addition to it.

But, Sir! you are, I am fure, far from
having any need to add that Motive: Or,
to put his Commands into the Scale, when
you well know of how much Weight
yours alone ever are with me. And tho',
if I confider how great his Penetration,
and yours is, I might be deterr'd from
offering any Thing I am able to write to
either, I am fo far encourag'd by your
joint Humanity, that without further
Hefitation, I venture freely to lay before both, what comes readily into my
Thoughts on the Subject He and You
think, and indeed very juftly, hath lain
hitherto fo much in the Dark.

The feveral Sorts of Matter, that
conftitute the terreftrial Part of the Globe
we inhabit, are ufually comprehended,
and

and fet forth by the Writers of Natural
Hiftory, under the general Name of
FOSSILS.

THESE are of two Sorts, *extraneous,*
and *native.* By *extraneous Foffils,* I intend the various *vegetable Bodies :* As
likewife the *Teeth* and *Bones* of *terreftrial Animals,* and the Shells of *Oyfters,
Conchæ, Cochleæ, Echini,* and other
marine Creatures, that are found in
great Numbers and Variety, buryed in all
Parts of the Earth. Thefe, by moft late
Authors, have been fuppofed to be found
in the Earth, and meer Stones; and treated of as fuch, under the Names of
Oftracites, Conchites, Cochlites, and
Echinites ; which Names occurr very
frequently in the Writers of Foffils.
And, by thofe Names, fometimes they
defign meerly the Shells above recited ;
fimple, free, and empty : Sometimes thofe
Shells fill'd with Stony, Flinty, or other
like Matter : Sometimes only the Stone,
Flint, Spar, or other Mineral Bodies,
that were originally formed and moulded
in thofe Sorts of Shells, fince perifhed

I   and

and gone : Sometimes the meer Impreffi-
ons of them in Stone : And not uncom-
monly, all thefe promifcuoufly and indifferently.   Which want of Care, and due
Examination of thefe fo different Bodies
was indeed one great Caufe that thofe
Writers fell into that Opinion.   But the
feveral *Sorts* of them are now rightly diftinguifh'd [a], and the *Origin* of each
afcertain'd [b].

I fhall only add here, for the further
clearing up of this Matter, the feveral
fanciful Names that have been heretofore
given to fome of the moft remarkable of
thefe Bodies : And, from my own Obfervations upon them, note what they really
are.   That commonly call'd *Cornu Am-
monis* [c] owes its Form to a turbinated
Shell : The *Bucardites* [d], to a *Bivalve.*
Indeed both of them are frequently found
actually covered with the very Shells in
which they were formed.   That Body to
K   which

[a] *Catalogue of the Foffils of* England, *&c.* M. S.
[b] Nat. Hift. *of the*
Earth. P*art* 4 *and* 5.
[c] Plin. xxxvii. 10.
[d] Plin. ibid.

which Dr. *Plot* [e] has given the Name of *Thrichites*, is affuredly only Part of the Shell, of the *Pinna-Kind*, compofed of tranfverfe Parallel Fibres not unlike *Hairs* [f], which was the Reafon that he confer'd that *Name* upon it. It is found very commonly, and in many Parts of *England*, befides *Shotover*, *Barton*, and the Places he mentions. The Figure of the *Hyfterolithus*, of which *Ol. Wormius* [g], and feveral Naturalifts fince, have imagin'd fuch ftrange Things, is wholly accidental, and taken from a Species of that Kind of Shell to which *Fab. Calumna* [h] has given the Name of *Concha anomia*; the Infide or Cavity of which this Stone is caft in, and exactly refembles. The *Brontia*, and *Ombria*, of *Geo. Agricola* [i], is an *Echinites*, and form'd in the Shell of the galeated *Echinus Spatagus*. So likewife are thofe of *J. de Laet* [k]; which he fuppofes to be alfo the

Che-

---

[e] *Nat. Hift. Oxfordfh. c.* 5. §. 145. *Tab.* vii. *Fig.* 7. *& Nat. Hift. Staffordfh. c.* 5. §. 40.

[f] Θρίκιτς, *Hairs.*

[g] *Mufeum. p* 83.

[h] *De Gloffopetra.*

[i] *De Nat. Foffilium. L.* 5.

[k] *De Gem. L.* 2. *c* 25.

Chelonites of *Pliny.* Thofe two grav'd
by *Fr. Lachmund* [1] are Stones form'd in
a different Species of the *Echinus Spatagus.* That which *J. Kentman* fent to
C. Gefner, whereof he has given an
Icon [m], is a Stone moulded in the Shell
of an *Echinus Ovarius.* He has alfo the
Figures of two Foffil-Shells of the *Echinus Ovarius* [n], fill'd with Stone. Thefe
he takes to be of that Sort that *Pliny*
calls *Ovum Anguinum.* The very fame
Ol. Wormius has caus'd to be engrav'd
under the Title of *Brontia* or *Ombria* [o].
Thefe Kinds of Stones the Country People here in *England* call fometimes *Fairy
Stones,* but commonly *Thunder Stones*;
in which Fancy they agree with the
People of *Germany* [p], and likewife
with *Pliny* [q]. The Bodies call'd *Tecoli-
thi* by *Pliny, Lapides Judaici,* and *Sy-
riaci,* by other Writers, fo much celebrated

[1] *De Foffil. Heldefhem.*
p. 3.
[m] *De Figuris Lapid.*
p. 61.
[n] *ibid. p.* 168.
[o] *Mufæum L.* 1. §. 2.
c. 12.

[p] Carmina ex eo no-
men invenit qnod cum
Fulmine, ut credit vul-
gus cadit. *G. Agrico. de
Nat. Foff. L.* 5.
[q] *L.* 37. *c.* 10.

brated by the antient Phyſicians for their
diuretic Properties, but reputed by all
as no other than meer Stones, have been
at laſt publickly demonſtrated [r] to be only
elevated Spikes of *Echini Ovarii,* brought
forth of the Sea at the Deluge, and buried, together with other marine Bodies,
in earth. The *Trochi, Trochitæ,* and
Entrochi, as alſo the *Aſteriæ,* are now
finally known to be all likewiſe owing to the Sea [ſ]. All the ſeveral Kinds
of each ſerve as Cords or Strings to tie
the Train or *Cauda* of that ſurprizingly
ſtrange Body the *Stachyoides* [t] to the
Shell of the Fiſh to which it belongs, and
ſerves as a Train and Rudder for Steer-
age. This *Train* has its Name *Stachyoides,* from its reſemblance of an *Ear* of
Maize, or *Indian Corn.* 'Tis found
commonly among Shells, and other Remains of the Sea, in ſeveral Parts of *Germany:* And Mr. *Roſinus* of *Munden* has
ſet forth a Diſcourſe [u] concerning it. I
have frequently met with *Parts* of it in
England,

---

[r] Greſham *Lecture, read*
May 9. 1693.
   [ſ] Greſh. *Lecture* 1693.

[t] Σ]άχυς. *Spica.*
   [u] De Stellis marinis,
Foſſil. 4°. *Hamb.* 4. 1719.

England, chiefly in the Chalk-Pits in *Surrey* and *Kent.* Mr. *Rosinus* calls the *Stachyoides,* Fossil *Sea-Stars,* I confess I cannot imagin for what Reason. The *Parts* and *Segments* of these Bodies have obtain'd various Names among the Writers of *Fossils,* e. gr. *Eucrinos, Penta-crinos, Pentagonos* [x.] The Bodies call'd by Mr. *Lhwyd Stellariæ* [y], are no other than Parts of the *Stella Arborescens.* The *Glossopetræ* are Teeth chiefly of *Sharks* of various Kinds. The *Plectro-nita,* or *Rostrage* of Mr. *Lhwyd* [z], is the Tooth of a strange Sea Fish, not nam'd nor describ'd by the Writers of Fishes. There is in my Collection, a Jaw of this Fish digg'd up, with Teeth of this Kind still actually remaining in it. The *Bufo-nitæ* are Teeth of the *Wolf-Fish* digg'd up in many Countries, along with other Spolia of the Sea. These were wont formerly to be worn in Rings, and pretended to have grown in the Heads of *Toads,* whence they had the Name of *Bufonii,*

---

[x] Lachmund. de Fossil. Hildeshem. *Sect.* 3. *c.* 17. 18.

[y] *Lithol. Brit. Tab.* 14.

[z] Lithophylacium Britan. *Tab.* 16.

Bufonii, and great Virtues afcrib'd to them. Dr. *Merret* [a], comparing thefe with thofe in the Jaws of that Fifh, found an exaƈt Agreement betwixt them, and rightly concludes both to be of the fame Origin. By this Method he imagin'd he had made a Difcovery of a Counterfeit and Impofture of the *Lapidaries* in felling *thefe Teeth* for the true *Toad-Stones*; fufpeƈting them to be really taken forth of the Jaws of that Fifh, and not out of the Heads of Toads, he feeming not to have known that there are naturally no fuch Stones in the Heads of Toads, that thefe are really all of them Teeth of the *Wolf-Fifh,* tho' thus found in the Earth; and therefore, by thofe who know not how they came there, reputed natural Stones. The *Siliquaftrum* [b] is evidently a bony Subftance, and by its Shape and Make appears to have ferv'd for Cover-ture and Guard of the Palate of fome Fifh, that feeds, as feveral do, upon Shell-Fifh.

[a] Pinax Rerum Nat. Britan. *p.* 210.
[b] *Mr.* Lhwyd *Philof. Tranf.* N°. 200. *and Litho-phyl. Brit. p.* 73.

Fifh. The *Icthyofpendyla* [c] are only ver-
tebres or Joints of the Back-Bone of
Sharks, and other Fifhes. The *Turcois*,
that hath paffed currently thorow all
Ages for a meer Stone, is indubitably of
Animal Origin. The various Samples of
it that I have feen, are fome of them
Fragments of very firm hard Bones, the
reft of Teeth, that have imbib'd a Tincture in the Earth, either a dusky Blue,
or a greenifh. The Teeth of various
Kinds of Sea-Fifhes, and of amphibious
Creatures, as the *Rofmari,* or *Morfe,* the
Manati, and of *Elephants,* left at Land
at the Deluge, are fometimes found in
digging, both here and Abroad ; of which
I have various Samples in my Collections. They are nearly of their Native
Complexion, where they have not been
lodg'd among mineral Matter, that being
infinuated into them has fuperinduced
and imparted to them its own Colour.
Thofe lodg'd where there is *Copper* in
the Earth, are frequently *blue* or *green,*
which Colour that Metal is wont to impart,

, Lhwyd Lithophyl. *Tab.* 18.

part, when infinuated in due Quantity.
Nay, even when in lefs Quantity, fo
that the Body is of its native pale Hue,
if expofed to the Heat of a Fire, to fetch
forth the latent Copper Particles, it becomes of a flight *Blue*, or a *Green*. To
the Bones and Teeth digg'd up out of the
Earth, that retain'd much their native
white Colour, or were a little variegated
with Black, which all the Foffil Elephants
Teeth, that I have feen, are, the antient
Naturalifts gave the common Name of
Ελεφας ὁ οςυκ7ός [d], *Ebur Foffile*. To thofe
that had acquired in any Part a bluifh
Colour, they gave the Name of *Calais;*
which, as fhall be fhewn by and by, is
what the later Writers call the *Turcois.*
Dr. *Poterius* [e], finding out that the *Tur-*
cois, which *Signior Pozzo* fhew'd him at
Rome, were really Ivory, tho' difguis'd
by the Colour, fufpefted them to be, becaufe not of real ftony Conftitution, all
counterfeit; which was the very Over-
fight that Dr. *Merret* fell into in relation
to the Bufonites.   As thefe Teeth and
Bones

[d] Theophraft. de Lapid.   [e] Pharm. Spagyr. *L.* 2.
c. 25.

Bones acquire a Colour by a *long* Stay in *cupreous Earth*, they attain it in a much *fhorter*, by their lying in *cupreous Water* ; this ferving *quickly* to introduce the metallic Corpufcles. Such there are in my Colle&ion, taken out of the Currents of Water that flow forth of the *Copper Mines* of *Herugrundt* in *Hungary*, and of *Goldfcalp* in *Cumberland*. Tho' *F. Hardouin* [f] doubts of that, *Salmafius* [g], and *Johan de Laet* [h], who had both of them much better confider'd, and been more converfant with Foffils than that learned Critick and Commentator, take the *Callais* of the Antients for our *Turcois* ; and, I think, with very good Reafon. *Plin.* L. 37. c. 33. *Callais e viridi pallens, fiftulofa, & fordium plena, ——leviter adhærens, nec ut agnata Petris, fed ut appofita, ——fragilis. Optimus color fmaragdi.* 'Tis not poffible any Defcription fhould better anfwer the Turcois ; which being a Tooth or Bone, that has lain long in the Earth, muft
L needs

[f] *Not. in Plin.* xxxvii. 33.
[g] *Exerc. ad Solin.*
[h] *De Lapid. & Gem.* L. i. c. 25.

needs be fofter and more *brittle* than real Stones, as alfo *foul,* as being fomewhat *porous,* which Teeth and Bones naturally are. Nor can it be *united, and of a Piece with the Rock,* wherein 'tis only lodged, but *flightly adhering* to it. Then the *Callais* was found in the fame Places, where we find our *Turcois.* As to the *Colour, Pliny* reprefents it here like that of the *Emerald*; by which *Cæfalpinus* fhews [i] he means a *Sky-Colour,* or *Blue-grey.* *Pliny* elfewhere [k] reprefents the *Callais* as nearly approaching the *Sapphire,* but paler, and of a *Sea-Green*; which exactly fuits the *Turcois.* And *Salmafius* well obferves, that the very *Name* fhews it to be of a *purpleifh* [l], or blue Colour. The *Hammites,* compofed ufually of multitudes of fmall globular Bodies, is wholely made up of a *Congeries* of the *Veficulæ* of the *Ova* of various Kinds of *Fifhes,* fill'd with a fine hard arenaceous Subftance. That they refem-
bled

[i] *De Metallis L.* 2.

[k] *Nat. Hift.* xxxvii. 56. Callais Sapphirum imitatur, candidior, & litorofo mari fimilis.

[l] Καλαϊον *Exerc. in Solin.*

I

bled thofe *Ova,* was indeed very early taken notice of ᵐ.

THOSE, which I have been hitherto difplaying before you, Sir! are the chief Particulars, I would note to you relating to the *extraneous Fossils :* And as to the *Native,* the Writers having been fo little accurate as, you fee, to confound Bodies of fo very different Origin and Conftitution with them, it cannot be thought ftrange, that their Accounts of the *native Fossils* themfelves fhould be frequently erroneous and imperfect. In affigning their very Names, they give us commonly the fame Body under different Names; as they do different ones under the fame Name. Then in their Methodizing and ranging of the native Foffils, 'tis no wonder that they fail, and that all Things are in Diforder, and out of Courfe with them, when they fo frequently make Choice of Characters, to rank them by, that are wholely accidental, and unphilofophical ; as having no Founda-

L 2   tion

ᵐ Hammites Ovis Pifcium fimilis eft, *Plin. Nat. Hift. L.* 37. *c.* 10.

tion in Nature, or the Conſtitution of the
Bodies themſelves. Thus ſome rank
them under the Heads of *common*, and
rare, of *mean* and *pretious* . of leſs, and
of greater *Uſe*. Then they reduce them
to *ſubordinate Claſſes*, according to their
particular *Uſes*, in Medicine, Surgery,
Painting, Smithery, and the like; which
would be proper in an Hiſtory of *Arts*,
or *Mechanics*; but ſerves only to miſ-
lead them and their Readers in the Hiſtory of *Nature*. Beſides, they rank, a-
mongſt the reſt, Bodies that are Mineral
indeed, but *factitious*, and not in their
native Condition. An Inſtance of this
we have in the *Pumex*, which almoſt all
the Writers of *Stones* place amongſt
them; whereas 'tis in Reality nothing but
a *Slag* or *Cinder*, found either where
Forges of Metalls have antiently been ;
or near *Ætna*, *Veſuvius*, or ſome other
burning Mountain, forth of which it
has been caſt. Another Example of this
we have in the *Lapis Spongiæ*, which is
a light, poroſe, friable Body, compos'd of
a Matter chiefly *Corallin*, and generally
made

made into the Form we find it, by a *marine Inſect.*

BUT theſe are only a few of the many Inſtances that might be alledged to evince in how uncertain and perplex'd a Condition this Study has hitherto lain : And how little Light into the Nature of Foſ-ſils, and their Relation to one another, we are to expect from thoſe that have heretofore wrote. The claſſical Diſpoſal of the native Foſſils will indeed ever be a Work of Difficulty. It hath been prov'd from Obſervations [n], made on the preſent Condition of them, that they have been once all in a State of Solution and Diſorder : And ſuch is the preſent Conſtitution of them that it is very hard, if not impracticable, to rank and reduce them into an exact Method [o]. For they want thoſe fix'd Characters of Affinity or Diſagreement that Animals, and that Vegetables carry along with them. It hath been ſhewn, how little Certainty there is in their Colour and Figure, in their

[n] *Nat. Hiſt. Earth, Part* 2.
[o] *Vid. Nat. Hiſt. Earth, Part* 4. ſub initio.

their Situation in the Earth, and their
Mixtures with each other P. And few of
them being pure, or unmix'd, 'tis plain
there can be no determinate Rule as to
their *fpecific Gravity*, their *Confiftence,*
or Approach more or lefs to *Solidity*, or
as to their Conftitution. In fine, there
being no fingle Character fteady, or to
be rely'd upon, I am oblig'd to make Ufe
of one or other of them, as I fee moft
fit, and conducing to my Purpofe. My
chief Regard is, to the Nature and *conftituent Matter* of each ; but fince that
Matter is frequently mix'd, and various
in the fame Sort of Body, I conduct my
felf by fuch other *natural Notes* as prefent themfelves, and all fuch *Tefts* and
Methods of Scrutiny, as I find practica-
ble. In particular, I have Regard to
the *Bulk* each Sort of Foffil is naturally
of: Alfo to its comparative *Gravity,
Denfity, Solidity,* the *Groffnefs, or
Finenefs* of the Parts : The natural *Figure* of the form'd Stones, and other Bodies, their *Texture* and *Conftitution ;* as
likewife

P Ibid.

likewiſe the *Colours* obſervable in many
Sorts of Foſſils, the *Diaphaneity,* or
Opakeneſs: Their Diſpoſition to a *Solution* and Mixture with Water. Laſtly, I
conſider in what Manner they affect the
Organs of Senſe, the *Smell* and the *Taſte;*
as alſo the *Touch,* as to their Roughneſs,
Harſhneſs, Smoothneſs, and their being
unctuous, oyly, and the like. With this
Conduct, and aſſiſted by theſe Lights, I
range the native Foſſils in the following
Method. 1. *Earths.* 2. *Stones.* 3. *Salts.*
4. *Bitumens.* 5. *Minerals,* or Bodies
nearly approaching the Nature of Metalls. And, 6. *Metalls* themſelves. The
particular Reaſons for my adjuſting them
thus, you will be better Judge of, when
you come to ſee the Detail of the whole
Method.

I am, Sir, *&c.*

LETTER III.

To the fame.

Of the Ceraunia, *or* Stone-Weapons, *the* Magical Gemms, *and fome other* artificial Things *antiently in Ufe, imagin'd by many late Writers to be* natural: *With* Icons *of feveral of thofe in my Collection, brought from moft Parts of the known World.*

SIR,

IT muft be allow'd, that I had the more Reafon to attempt the *Natural Hiftory of the Earth,* and of the Bodies found in it, both *native* and *extraneous,* becaufe, as you obferve, this Study had all along lain in the greateft Darknefs and Confufion: And, to the very Time that I fet forth that Work, it was *not yet agreed among the Learned, whether thefe Bodies formerly call'd* Petrify'd Shells, *but now-a-days paffing by the Names of formed* Stones,

Stones, be original Produttions of Na-
ture, form'd in Imitation of the Shells
of Fifhes, or the Shells themfelves p.
Indeed the lateft Writers of all were pofi-
tive that thefe Bodies were not *real.* Dr.
Lifter q afferts point blank they were *ne-*
ver any Part of an Animal, being only
Refemblances of Shells, but *meer Stones,*
which the *Earth produces,* and each
fhap'd by the Power inherent in the Stone,
or in it felf. This muft needs be allowed
by all who have made any Obfervations
of the Produttions of Nature in the Formation of Bodies, tho' they have not
made many Obfervations on thefe, to be
a Doftrine, however pofitively delivered,
very myfterious and paradoxical. Be that
as it will, not only Dr. *Lifter,* Dr. *Plot,*
and others here, but learned Men A-
broad, fell generally into it. Nay, fo
Zealous were they bent upon it, and
ftrongly poffeffed with it, as to imagin
not only the animal and vegetable Bodies,
found lodg'd in the Earth, but feveral

M

artificial

p *Mr.* Ray's 3 *Difcourfes* 8o. Lond. 1693. *p.* 127.
q *Philof.* `Tranf.` No. 76. *Cont. Lib.* Cochlitarum
Angliæ 4o.

artificial Things, antient *Urns*, and other *Vafes, Stone-Weapons,* and *Magical-Gemms,* to be productions of it, and formed by Nature under Ground; which may pafs for one of the many furprizing Inftances there are of Precipitation, Credulity, and want of Judgment in thefe Writers; and I wifh there were not fo many likewife in all the other Parts of natural Hiftory; that a Man that would be accurate in any, can hardly tell what to rely upon, without bringing all, of the very much that hath been written, to the Teft anew. I have formerly had Occafion to make fome Reflections on the Notion [r] of the *Foffil-Urns*; and fince I have your Commands for it, I fhall here offer fomething concerning the Antient *Magical Gemms,* and *Stone-Weapons.*

Dr. *Lifter* [f] fuppofes thefe Gemms to be *Ombriæ;* and with his ufual Warmth and Pofitivenefs, pronounces them *figur'd*

[r] *Nat. Hift.* Telluris defenfa contra Camerar. *p.* and Mr. Holloway's *Tranflation, p.* 154.
[f] Philof. Tranf. No. 201.

I

gur'd naturally, and without any Arti-
fice: Nay, and which is a very pretty
Fancy, *naturally polifh'd too,* with juft
as much Reafon as he might a *Table Dia-*
mond, a *Brillant,* or an *Intaglia* of *Ju-*
lius Cæfar. But you know, Sir! this
learned Gentleman having fet forth in his
Youth, with the Notion, that all Bodies
of regular and determinate Figure, found
in the Earth, were form'd there, abid
by it ftifly to the End; this being the
very laft Paper he publifh'd on this Subject. Writers for Fame, great Souls!
are ever conftantly in the Right, and will
fooner give up their Lives than their Opinions; even tho' they firft take them up
frequently upon meer Fancy, or very
flight Grounds; while thofe, who really
fearch after *Truth,* are very wary in
what they advance; and with great Rea-
dinefs and Candor fubmit all to the
ftricteft Scrutiny, attending as well to
every Thing that may be offer'd againft
it, as for it. As to the Bodies you are fo
defirous of an Account of, they have
pafs'd from the remoteft Antiquity downwards, under the Notion and Title of

Magi-

Magical-Stones, or *Gemms.* They are, to this Day, fometimes found broad in our fields. I have feen only three Kinds of them, and keep a fair Sample of each in my Collection. Neither any other Writer, nor Dr. *Lifter,* mentions any more: And his are indeed the fame with mine; fo that I am pt to believe there are no more. My firft is, of an exactly fphærical Form, near two Inches in Diameter. The fecond is a *Sphæroid,* much comprefs'd, 1 Inch $\frac{4}{10}$ in Diameter, and $\frac{7}{10}$ in Perpendicular. The third is oblong, round off at each end, with a Bafis fomewhat convex, and two Sides alfo a little fwelling and convex, the upper Part terminating in a Ridge. This Stone is two inches in Length, and $1\frac{3}{10}$ in Diameter. 'Twas found near *Barkhamftead* in *Middlefex,* and poffefs'd long, indeed to his Death, by an eminent Phyfician there. 'Twas made ufe of by him as a *magical Speculum* ; he giving out to his Patients, that *Something* was wont to *difcover* it felf to him in *this Stone,* by which he receiv'd Light and Informations, on fuch Occafions as he infpected and confulted it.

He

He left a great *Estate* to his *Son*; who not being ever able, with both his Eyes, to discover that *Spectrum*; instead of getting an *Estate*, spent the greatest Part of that which his Father left him; and was pleas'd to do me the Honour to send the Stone to me, who being not so happy as to be possess'd of *Faculties* equal to those of the wise good *old Gentleman*, can no more discern the *Spectrum,* nor get an *Estate* by it, than the generous frank young Gentleman his Son could.

These three Stones are all form'd out of that Sort that the Lapidaries call *Peble-Crystal*; which is found in several Parts of *England*; and are very fair, pellucid, and clear. The first is indeed of a fine deep Water, and is a very beautiful Stone ; being of a sphærical Figure, it might be taken for a *Pearl:* And Dr. *Lister* [t] says, that these are *call'd in some antient Leases, Mineral-Pearl.* In former Times, they must of Course be, before

Ibid.

fore they were pick'd up, more frequent,
particularly in *Britain :* And 'tis not altogether improbable, that thefe are of
thofe mention'd by *Suetonius* [u] as found
antiently here, and fuppos'd by the *Romans* to be *Pearls,* but of an *extraordinary Bignefs* [x]; thefe being indeed vaftly
more large than any of the true *Pearls.*
moft of thefe Stones, and particularly the
three above mention'd, are fo regularly
cut, and polifhed in a manner fo exqui-
fite, that I can hardly imagine how a
people fo barbarous, and deftitute of all
Working-Tools, [y] could ever finifh them
with fo great Elegance and Exactnefs.
When firft I obferv'd thefe Stones, I con-
jectur'd they might be us'd meerly as
Gemms, and worn antiently for Ornament by the Natives.   But Mr. *Aubrey,*
who, you know Sir! hath much ftudyed
the Antiquities of this Ifland, contends
that they were us'd in *Magick* by the
Druids : And, in his *Mifcellanies* [z], he
takes notice of a *Cryftal Sphære,* fuch as
the firft of thefe is, or *mineral Pearl,*
us'd

---

[u] *In Cæfare* §. 47. [x] Sueton. Ibid. [y] *Conf. State of*
London, 8º. §.—— [z] 8º. Lond. 1696. Page 128.

us'd by Magicians, and *to be infpected by a Boy.* But, long before him, *Joach. Camerarius* [a] *mentions a round Cryftallin Gemm,* into which a *chafte Boy* looking, difcern'd an *Apparition,* that fhew'd him any thing that was required or fought for. *Paracelfus* [b] carries the Thing further, and avers, that in thefe *fpecula are feen Things paft, prefent, and to come :* And that fome *Star impreffes on the Cry-ftal an Image of its Influence, and a Simili-tude of the Thing inquired and look'd for in it.* And of this Sort were the Cryftal-lin Stones made ufe of by Dr. *Dee,* and Mr. *Kelly* in their myfterious Vifions and Operations ; of which they drew up a *Journal,* fince publifh'd by Dr. *Meric Caufabon* [c]. One of theirs was *round, of a pretty Bignefs,* and of *Cryftal* ; very probably the fame with my firft. This they call the *Shew-Stone, and Holy-Stone.* You fee, Sir ! from thefe Foole-ries having held and been kept up thus, from the moft early Times, in a continued

[a] *Præf. in* Plutarch. de Defectu Orac.
[b] *Explic. Aftron.*
[c] *Relation of Dr.* Dee, *&c. Fol.* London 1659.

nued Tradition, quite down to our own, while Things the moft highly rational have been neglected, dropp'd and fallen into Difufe, how fond Mankind hath ever been addicted and prone to Superftition ; of which there are but too many other Inftances.

A s to the Antients, from the Writers of thofe Times, we learn, that the *Zoranifcos* was a *Gemm* us'd by the *Magi* [d]; as alfo the *Heliotropium* [e]; with a great Number of others, not needful to be recounted here. Thofe which come the neareft to ours, and from which the fuperftitious Ufe of thefe feems to be derived, are of the *Star-Stone,* or Aftroite Kinds. Not of thofe of the later Naturalifts, which had their Names from their Figure, or fome Delineations upon them, refembling the Stars of Painters and Heralds, but of thofe of the Ancients which were lucid and tranfparent ; and therefore were faid to *fhine like a Star,* whence they had their Name.

—Καλα

---

[d] Zoranifcos Magorum Gemma. *Plin.* xxxvii. 10.
[e] Ibid. xxii. 29.

—Καῖὰ σκοπιὰς Παλλήνης
φύεῖαι 'Ασέριος, καλὸς λίθος, ὃια ῖις 'Ασηρ
Μαρμαῖρων [f].

I n like Manner the *Star-Stone* of
Pliny was white, or nearly approaching
the Tranſparency of *Cryſtal*, and ſuppo-
ſed to have its Name from *reflecting back
the Light of a Star, when expoſed to it* [g].
The ſame Author treating, if not of this,
of a nearly related Species of *Star-Stone*,
which he ranks likewiſe amongſt the
tranſparent Gemms, tells us that 'twas
in *mighty Eſteem*, and that *Zoroaſter*, one
of the moſt celebrated of all the *oriental
Magi, had ſet forth its wonderful Effi-
cacy* in *magical Arts* [h]. The ſame Author obſerves, that the *Aſteria* was a
pretty hard Stone, and that the Lapidaries found ſome Difficulty in the cutting

N   of

[f] Dionyſ. περιηγ. 327.
[g] Candida eſt vocatur
Aſterios, Cryſtallo pro-
pinquans, in India naſ-
cens, & in Pallencs Lit-
toribus— Cauſam Nomi-
nis reddunt quod Aſtris
oppoſita Fulgorem rapiat
ac regerat. *Plin.* xxxvii. 9.
[h] Celebrant & Aſtroi-
tem, miraſq; Laudes ejus
in Magicis Artibus Zoro-
aſtrem ceciniſſe. *Plin.*
xxxvii. 49.

of it; which is likewife the Cafe of thefe
Stones. So that thofe of the Antients
apparently agree with thefe as to their
Conftitution, their Complexion and Diaphaneity, as well as the fuperftitious
Ufes they were applied to: And I take
notice of one Species of *Star-Stone* in
the fame Author, that was likewife *or-
bicular*, and of the very fame Shape
[i] with the firft of mine. I'm not a little
difpleas'd with my felf, that I have, before I was aware, taken up fo much of
that Time which, Sir, you know fo
well how to imploy better, and run
on thus far on a Subject fo very flight.
But I fhall difmifs it, after I have offer'd
you a Conjecture at the Reafon why this
Kind of Stone has been employ'd thus as
a *Speculum*, and turn'd to Magical Delu-
fion, and the fpying out of *Spectra*. It
moft probably happen'd from the *Conftitution* of the Stone ; which, in every various Pofition, gives a various Corufcati-
on, and Glare of the Light ; and, by that
Means, a various Reprefentation of
Things

Sideritis— Globofa Specie. *Plin.* xxxvii. 65.

Things, and Entertainment of the Fancy.
Which Conjecture I am led into by their
own Deſcriptions and Accounts; wherein they ſet forth the *Glittering and Light*
of the Star-Stone, which they compare
ſometimes to that of the *Sight of the Eye*,
ſometimes to *the Moon at full*; and take
notice beſides, of its reflecting the *Light*
of a *Star*, or of the *Sun*, when expoſed
to either [k].

IN like Manner the *Selenites*, or
Moon-Stone of the *Antients* was *white*,
or *tranſparent* [l], and had its Name from
repreſenting the *Moon*, as in a Glaſs, as
Pliny, *Geſner* [m], *Agricola* [n], and Dr.
Plot [o] obſerve; tho', for the ſame Rea-
ſon, it might as well have been call'd the
Sun-Stone, it as readily repreſenting that,
or any other luminous Body; and there-
fore had likewiſe the Name of *Lapis-
Specularis*, as Dr. *Plot* takes Notice.
And as the *Aſtroites* was uſed in Ma-

N 2　　　　gick,

---

[k] *Plin. L.* 37.
[l] *Plin. ibid. & Dioſcorid.* ὕλης ἰατρ. v. 159.
[m] *De Fig. Lapid.*
　De Nat. Foſſil. L. 5.
　Nat. Hiſt. Oxf. c. 5.

gick, amongſt the *Antients*, ſo the *Selenites* was uſed by them as a *Charm* or
Amulet ᴾ.

You'll imagine, Sir! treating of theſe
Things, 'twill not be eaſy for me, not to
recollect the ſo juſtly celebrated and il-
luſtrious *Oracle* of the *Jewiſh Nation*,
that paſs'd among them, under the Name
of *Urim and Thummim,* or *Lights* and
Perfections, for ſuch thoſe Words import.
This was compoſed of *twelve Gemms,*
artfully join'd together, and worn on the
High Prieſts *Pectorale*. 'Tis thought,
whether rightly or not I take not upon
me here to determine, by the beſt Judges
of the *Jewiſh* Antiquities, that thoſe
who conſulted this *Oracle*, looking in-
tenſly upon it, receiv'd Anſwers and Re-
ſolves by ſome new and unuſual *Lights*
and Irradiations then miraculouſly exer-
ted and caſt forth by thoſe *Gemms*. The
Fame of a Thing ſo ſurprizing and
extraordinary could not but paſs Abroad
to the neighbouring Oriental Nations ;
                                    and

[ᴾ Φυλακτήϱιον, πϱείαμμα. Dioſcorid. *I. c.*

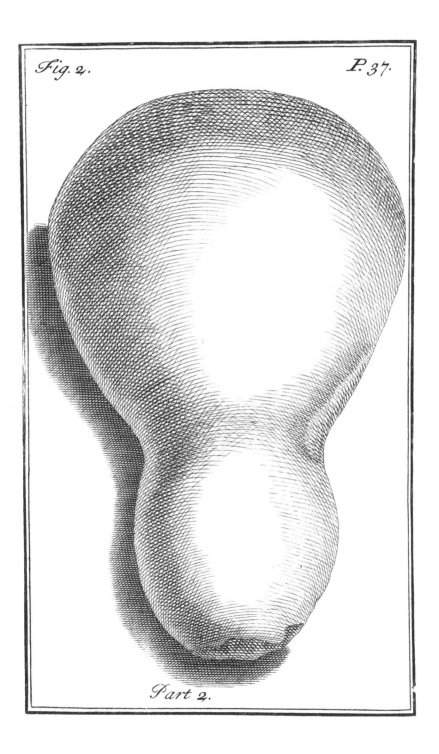

Fig. 2.

P. 37.

Part 2.

Fig. 3.

P. 37.

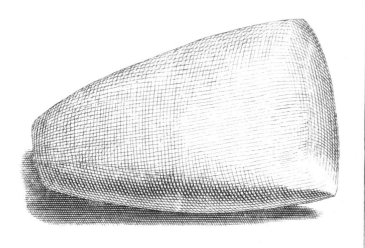

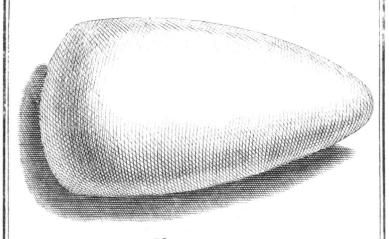

Part 2.

Fig. 4.

P. 37.

Part 2.

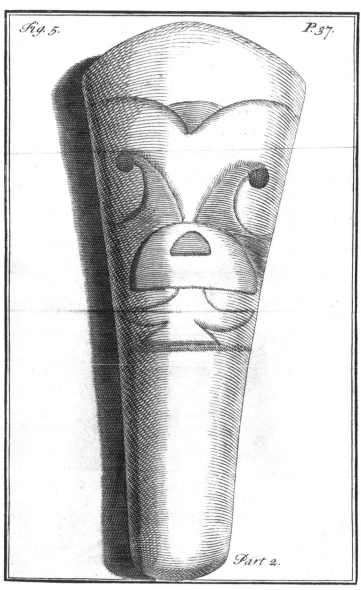

Fig. 5.   *P. 37.*

Part 2.

The material originally positioned here is too large for reproduction in this reissue. A PDF can be downloaded from the web address given on page iv of this book, by clicking on 'Resources Available'.

and 'tis not wholly improbable, that the
Zoroaftrian, and other like *Gemms*,
were made in Imitation of this, and took
their firft Rife from it. 'Twill not be
thought Strange, that they fhould all
differ fo much from this, when 'tis known
that the *Jews* treated all the other Nations in a manner very fupercilious, and
were fhy of imparting any Thing to
them ; fo that the Tradition and Account
they receiv'd of it muft needs be very
lame and imperfect.

I come next, Sir ! in Purfuit of your
Commands, to give fome Account of the
Stone-Weapons and *Inftruments*. Now,
tho' thefe carry in them fo plain Tokens
of *Art*, and their Shapes be fuch as apparently to point forth, to any Man that
rightly confiders them, the *Ufe* each was
deftin'd to; yet fome of the Writers of
Foffils, and of great Name too, have
been fo fanguine and hafty, fo much blinded by the Strength of their own Fancy, and prepoffeffed in Favour of their
Schemes and Notions, that they have fet
forth thefe Bodies as natural Productions
of

of the Earth, under the Names of *Ce-rauniæ*. Of this Sort are the *Ceraunie* of which we have *Icons* in *C. Gefner* [q], *A. Boetius* [r], *M. R. Befler* [f], *Ol. Wor-mius* [t], *S. L. Mofcardi* [u], and *Fr. Lach-mund* [x]. And *J. Kentman* [y] hath left us a Defcription of four of thefe, likewife, un-der the Names of *Ceraunie*. The Au-thors here recited, imagine thefe to be the *Ceraunie* of the *Antients*. Probably they may be thofe of *Sotacus* [z]; but what the *Ceraunie* of *Pliny* were, it is not eafy to conjecture from his Account of them [a] He fuppofes them to fall with *Showers* and *Thunder*. As he does like-wife

---

[q] *De Tig. Lapid. p.* 62. 64.

[r] *Hift. Gem. L.* 2. *c.* 261.

[f] *Gazophyl. Rer. Nat. Tab.* 34.

[t] *Mufæum. L.* 1. *Sect.* 2. *c.* 12.

[u] *Mufæo Mofcardo L.* 2. *c.* 50.

[x] *De Foffil. Hildefhem. p.* 23.

[y] *Nomenclat. Foffil. Mifniæ p.* 30.

[z] Sotacus & alia duo genera fecit Cerauniæ, ni-gras rubentefq; ac fimi-les eas effe *Securibus* ; per illas quæ nigræ funt & rotundæ Urbes expugnari & Claffes, eafque *Betulos* vocari : quæ vero longæ funt, *Ceraunias*, Plin. *L.* 37. *p* 737.

[a] Eft inter candidas & quæ *Ceraunea* vocatur, fulgorem fiderum rapiens. Ipfa Cryftallina fplendo-ris cœrulei. *Plin. L.* 37. *p.* 737.

wife the *Ombria,* and *Brontia* [b] : Of the *Ombria* he gives no Defcription, and a very obfcure one of the *Brontia*; he only comparing it to the *Head* of a *Tortoife* [c]; as he does alfo the *Chelonitis* [d].

T H E *Stone-Weapons,* and *Inftruments* were all cut out, and made, before the Difcovery of Iron. But, when once this Metal was brought to Light, and its *Ufes* known, 'twas found fo much pre-ferable in every Refpect, that thofe Stones were prefently caft away : And they are thofe which we ftill fometimes find A-broad in the Fields, not only here in *En-gland,* but in *Scotland* likewife, and *Ire-land,* and *Germany,* and feveral other Countries; where they ferv'd, in the moft early Ages, for Axes, Wedges, Chizels, Heads of Arrows, Darts, and Lances. Nay, among Nations yet barbarous, and unacquainted with the Manufacture of Iron, and that have not been difcover'd

by

---

[b] Ombria ficut Cerau-nia, & Brontia cadere cum Imbribus & Fulmi-nibus dicitur, *Plin. L.* 37. *c.* 10.

[c] Brontia capitibus Te-ftudinum fimilis *Plin. L.* 37.

[d] Chelonitides teftudi-num fimiles. *Plin. L.* 37.

by the *European* Navigations, till of late
Years, thefe *Stone-Weapons* and *Inftru-*
ments are in Ufe to this Day ; e. gr. in
the Ifland of *Guam*, one of the *Ladrones,*
and in *Nova-Britannia*, an Ifland lying
South of the *Æquator*, and the fartheft
Eaft of any yet known, difcover'd a few
Years ago by *Captain Dampier.* Indeed,
when the *Spaniards* made their firft De-
fcent upon *America*, they found no *other*
amongft any of all the Nations of that
vaft Continent, or the Iflands adjacent.
For, tho' the *Americans* had in many
Parts *Iron-Ore*, very good, and in great
Plenty, they knew not the Ufe of it, till
they were taught that by the Spaniards.
In my *Difcourfe of the Peopleing of*
America e, I have fhewn, that *that Colo-*
ny was departed, and had left the *old*
World before Iron was found out, and the
Ufes of it known there. They are fo
many and great, and this Metall of fuch
Importance, that, had the *American*
Colony been acquainted with it, before
their

e *Of this there is fince fet forth a brief* Extract. Nat.
Hift. *Earth illuftrated.* p. 105. *& Seq.*

their Departure, they would never have again loſt or forgot it. Perhaps, Sir! you may ſay, that there were *Iron In-ſtruments* in the World long before, even before the Deluge; which we learn from the Hiſtory of *Tubal Cain*, who was then *an Inſtructer of every Artificer in Braſs and Iron* [f]. Now theſe muſt be known to *Noah*, and all his Sons, by whom the *whole World* was-peopled. But thoſe *In-ſtruments* all periſh'd, and were deſtroy'd, during the *Deluge.* I have ſhewn elſe-where [g], that all *Metallic* and *Mineral Bodies* were then diſſolv'd: And, tho' it be there ſo clearly made out, from Obſer-vations, that none be ſtill wanting, this affords an *additional Proof* of the *Cer-tainty* of *that Propoſition.* From the moſt indubitably authentick Monuments that can be required, we know that the *Uſe of Iron* was not recovered in *Aſia*, whence it paſs'd to *Europe*, and the reſt of the *Old-World*, till ſome *Ages after the Deluge:* Nor in *America*, till the *Spaniards* made their Deſcent upon it, two or three Cen-

O   turies

Gen. iv. 22.   [g] *Nat. Hiſt. Earth. Part.* iv.

turies ago. And tho' *Noah* and his Sons
could not but remember the Iron in Ufe
before the Deluge; yet fo great Havock
and Devaftation was made, during that
fo fatal and terrible Cataftrophe, and fo
unhappy a Change in the Earth, that
there was every where a new Face of
Things, in which they muft be much to
feek, and reduced to the greateft Diftrefs,
Exigence and Neceffities [h]. They muft,
in fuch a State, be fully taken up meerly
in providing Food, and the common Sup-
ports of Life; and would have little leifure to look after Arts, and Things of re-
moter Ufe, till Mankind were further
multiplyed, and their Affairs on a better
Foot. In this fo calamitous a Condition,
Iron might be perfectly *forgot*, and the
Knowledge of it quite worn out.

'T is remarkable, Sir! that, of thefe
antient *Stone-Weapons* and *Inftruments,*
many are fhap'd with great *Regularity*
and *Art*, and `finifh'd with an Exactnefs
very furprizing, confidering they then
had

[h] *Conf. Nat. Hift. Earth. Part 1.*

Fig.6.

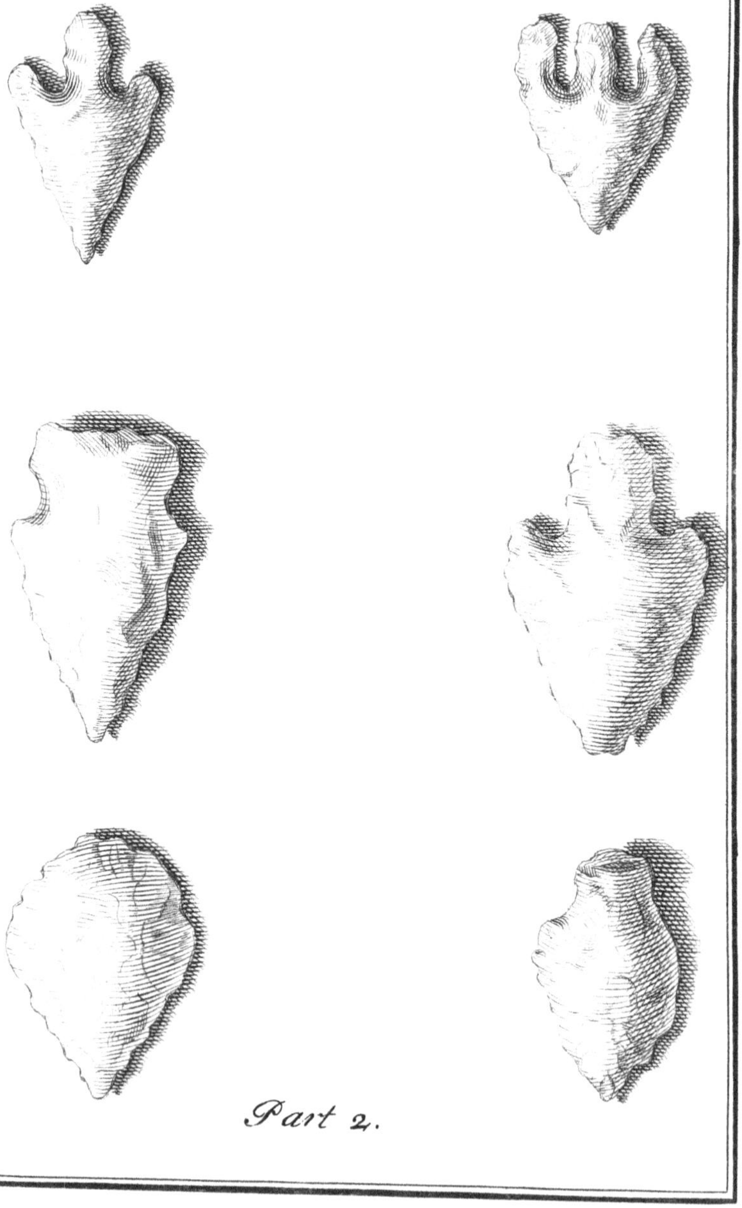

Part 2.

had not the Aſſiſtance and Advantage of the *Tools* that we now have. The *Arrow-heads* are particularly remarkable. They are of a Form the moſt miſchievous, and fitted to hurt, that could be poſſibly de-viſed. They are brought to an exquiſitely ſharp Point, keen Edges, and have Snaggs, or, as they are call'd, Beards, on each Side, on Purpoſe that they may make a large Wound wherever they en-ter, and not be drawn out again without much Difficulty and Harm to the Part in which they happen to be infixed. 'Tis further remarkable, that the *Arrow-heads*, found in Countries the moſt di-ſtant each from other, e. gr. *Britain*, and the Country bordering on the *Streights of Magellan*, are of the *ſame miſchie-vous Form*. 'Tis true, different Men having in View the ſame Deſign, condu-cting their Thoughts in a regular Man-ner, may come, in the Purſuit, to the ſame Concluſion ; and, as in this Caſe, hit on the ſame Shape for a Weapon of ſuch Deſign. But it is much more likely, that they came all from the ſame Origin ; and that the *firſt Module* was brought from

Babel,

Babel, to the various Countries whither the several Colonies, sent thence, made their Migrations.

Give me Leave, Sir! to take notice by the by, that some may perhaps think strange, that such a Building as that of *Babel* should ever be set about by Men that had not the Use and Affiftance of Iron But this, like the other most antient Buildings of those Parts, was of *Brick*; in which Iron Tools are not so needful as in Buildings of *Stone*. And yet I have Reason to believe, that the vastest of this later Kind that the World ever saw, I mean, the *great Pyramid* in *Egypt*, was rais'd before the Managery of that Metall was again recover'd and found out.   Be that as it will, 'tis most certain, that the Buildings, that the *Spaniards* found out at their first Arrival in *Peru*, were rais'd and finish'd wholly *without the Affiftance of Iron*.   And yet several of these were so magnificent, and some of the *Stones* so very *large*, as almost to amaze the *Spaniards*.   What added to the Surprize, was, that all were very *regularly wrought,*

wrought, ſquared, and the Joints ſo clo-
ſed and fitted, as hardly to be diſcern'd;
in ſo much that the whole appear'd as if
cut out of only one huge Maſs of Stone.

B y Means of a Multitude of Hands,
and united Strength, with continued Labour and Induſtry, a right Invention and
Contrivance, Things ſurprizingly great
have been perform'd even by Nations the
moſt barbarous, ſavage, and wholly de-
ſtitute of Inſtruments and Machines. As
to thoſe Buildings, 'tis probable the *Pe-*
ruvians ſquared and ſmoothed the Stones
by *rubbing* them againſt one another; and
raiſed them up into their Ranks and Places by Aſſiſtance of *Heaps of Sand* or
Earth, gradually piled up on the outſide
of the Walls of theſe Buildings.

'T i s late: And, which, I fear, you'll
but too eaſily collect from what, preſum-
ing on your uſual Indulgence, I thus venture to ſend you, I am very ſleepy ·
which falls out the more fortunately to
you, as it prevents your further Trouble.

I am, Sir, *&c.*

LET-

NUMBER IV.

An Extract *out of the* Preface *of one of the* Catalogues *of my* Foffils, *containing* Directions for regiftring of the native Foffils, and compofing an inftructive and ufeful Catalogue of them.

T HESE Foffils ought firft of all to be digefted into Claffes, and enter'd in a proper Series and *Method*, according to their mutual Relations and Alliances. Then the *Hiftory* of each fhould be given; fo far as there can be any Knowledge or Information of it obtain'd ; with an Account *where it was found*, at what *Depth* in the Earth, what *other Bodies* or *Matter* it was attended with, in *what Manner it lay*, whether in a *Fiffure*, or in a *Stratum* ; with all other Circumftances of the Place. Next fhould be noted every Thing obfervable in the Body it felf; its *Colour,* its *Figure, Texture,* or the Manner of the Concretion

of

of the Parts; and the *different Sorts of Matter* that concur, and are united in the fame Mafs. *Finally,* each fingle Body fhould be brought to the *Fire,* to *chymical Trials,* and all other Tefts; in order thorowly to difcover its *Nature, Conftitution, Properties,* and various *Ufes.* Was this once effectually done, and juft *Deductions* and *Inferences* made from the Whole, 'twould go a great Way towards a *Natural Hiftory of Foffils, and* the perfecting this Knowledge. Of the great Profit and Ufefulnefs of thefe Studies to the Publick, I have fpoke fully, and given many Inftances elfewhere. What adds further to their Advantage is, that they are not only *entertaining and pleafant,* but if the Compiler be accurate, they muft be *clear* likewife, fure, and little liable to Error and Impofition. *Mathematical Propofitions* are ordinarily abftracted; require great Extent of Thought, and Application of Mind. Whereas thefe *Mineral Propofitions* are plain, fimple, and obvious. The Relations of the Site and Circumftances of the Foffils in the Earth, and of the

various

I

various Experiments made upon them, are no other than fo many *Hiftories of Faƈt*. The Accounts of all Things obfervable in the Foffils themfelves, will carry with them *Evidence of Senfe,* which is the *higheſt Certainty*. Thefe Foffils will be fo many ftanding *Monuments*, that give *perpetual Atteſtation* to this : And there can need no other *Proof* of thofe *Accounts* than only fimple *View* of the *Things* fet forth in the *Catalogues*. Nor, finally, can it be difficult to difcern, whether the *Conclufions* drawn from thofe Relations, Experiments, and Accounts, *follow rightly* from them, or not.

N U M-

NUMBER V.

To Mr. ⸺

The Aſſiſtance that this, and ſeveral other learned Men have given to the carrying on the Deſign of the N. H. E.

Sir,

THERE are not many in this Age, who have taken the Pains that you have done, very happily and ſuccefsfully, in moſt Parts of uſeful Learning; but more particularly in the Study of the *Natural Hiſtory of the Earth*, and of *Foſſils*. The Example and Countenance of a Gentleman of your Diſtinction and right good Senſe, has been an additional Confirmation and Incitement of me; and the Communications that I have from Time to Time receiv'd from you, have given me no little Light and Aſſiſtance. Such Part of my Labours, as I have ſub-mitted to the Judgment of the Publick,

P   have

have met with greater Oppofition from
fome, they beft know why, than I had
Reafon to expect. But when I confider'd what it was that they urg'd, it rather afforded me Reafon to believe what
I was doing was right, and confirm'd me
in the Purfuit of it.

Tu ne cede malis, fed contra auden-
   tior ito.

I can eafily pafs by Opinionatry, Ill-na-
ture, and the bufy meddling of thofe
who thruft themfelves into every Thing,
how little Knowledge foever they may
have of any Thing, while I have the
Approbation of Men of your Candour
and Accomplifhment. Nothing can give
me higher Encouragement. 'Tis for the
Satisfaction of fuch only that I was concern'd : And, having attain'd that, I have
my End. What you write in your laft,
that having had Occafion feveral Times to
pafs and repafs the Alps, *where fuch vaft*
Tracts of the interior Parts of the Earth
are difplay'd, and laid open to view, and
various Opportunities for feveral Years
                                   paft,

*paſt, of making Obſervations in many
other Places, you are perfectly convinc'd
of the Truth of theſe Obſervations that I
have publiſh'd in my natural Hiſtory of
the Earth : And that, after having carefully conſider'd them, you are as fully ſa-
tisfyed in the Concluſions that I have
drawn thence : And that mine is the only
Hypotheſis that anſwers Nature, and
ſolves all the Phænomena obſervable in
the Earth, in an eaſy and Geometrical
Manner.* This, I ſay, keeps me in Countenance, and is a ſufficient Support to me
againſt other Gainſayers : And 'tis with
no little Satisfaction that I take notice to
you, that from what they print and de-
clare, 'tis evident, that the Impartial all
over *Europe* have the ſame Sentiments.
It muſt be allowed a fair Preſumption in
Favour of the Truth of my Doctrines,
that they have abid a very rigorous Teſt
now for above thirty Years, ſtand yet
firm; and the longer and more ſtrictly
they are look'd into, the more they are
confirm'd to this very Day. Give me
Leave to lay before you the Opinion of
one that is ſtill actually engag'd in theſe

Searches,

Searches, very curious, a good Judge,
and has carry'd them on over a great Part
of the Globe, from *Numidia*, along the
northern Parts of *Africa*, by the Ruins
of ol d *Carthage*, quite on to *Egypt*, to
Arabia, Phænicia, Syria and *Palestine;*
Countries from which we have hitherto
had very few Accounts. ·This is Mr.
Thomas Shaw, Fellow of *Queens College*
in *Oxford*. The Words of his Letter to
me, *June* 1. 1726, are, —— *Wherever*
I have been, I have had such convincing
Proofs of what you advance in your natu-
ral History of the Earth, that my Voy-
ages are only imperfect Comments, and
smaller Testimonies of what you have
elsewhere much better observed. I am
sure a Person of your Curiosity, will be
pleas'd to know one Particular, which
this ingenious Gentleman acquaints me
with in another of his Letters. As he
was making Observations upon the *great*
Pyramid, he took notice of *Shells*, and
other *Marine Bodies*, lodg'd in great Variety and Abundance, in the Mass of the
Stone, of which that Pyramid is built, and
in that of the Rock wherein it stands,
which

which is of the fame Sort, and indeed in
other Parts of the Country ; which was
obferv'd of the Mountains of *Egypt* 2000
Years ago by *Ariftotle*, and others of the
Antients, Now this *Pyramid* is one of
the firft *Structures* that was rais'd after
the *Deluge*. Indeed it was built within
250 Years of the Time of that great
Cataftrophe, when, you know, I have
afferted thofe Shells were brought forth
of the Sea, and repofited in the Strata of
Earth, and of the Sand, that afterwards
gradually hardned into Stone. Mr.
Shaw's Obfervation muft be allow'd a
confiderable Confirmation of my Do-
ctrine. The *Marine Bodies* in the Stone
of the *Pyramid*, carry the Thing to
near the Time I propofe : And thofe in
the Strata of the Rock underneath, quite
to it, and up to the very Time of the Com-
pilation of thofe Strata ; which was during the Deluge [a]. The learned and ingenious *Steno* [b] made a like Obfervation
in the vaft Stones in the *Ruins* of the
Walls of *Volaterre* in *Tufcany*. In thefe
he

[a] *N. H. E. Part* 2.
[b] De folido intra folidum. 4to *Flor.* 1669.

he found incorporated all Sorts of Sea
Shells ; which therefore muft have been
exiftent before the Time that thofe *Walls*
were built, which was feveral Centuries
before the Building of *Rome ;* and that
carries them back to within not many
Years of the Time of the *univerfal
Deluge.*

As you Sir ! and Mr. *Shaw,* fo fome
others of the greateft Men in *Europe,*
from the Time that my *Natural Hiftory
of the Earth* firft came forth, have done
me the Honour to affift me in the carrying on that Work, at their no fmall Pains
and Expenfe. Of thefe Dr. *Scheuchzer*
Profeffor of Mathematicks at *Zurich* is
one. You are well acquainted with his
Perfon, his Works, and his great Me-
rits. Dr. *Leopold* of *Lubeck* is another ; who finding my Colle&ion not
fufficiently ftored with *Swedifh Foffils,*
and that I had not a fatisfa&ory Account of the *Mines* there, of his own
Accord, and at his own Expenfe, un-
dertook a Journey thither for my Sa-
tisfa&ion ; with what Succefs you may
fee

fee in a Letter that he was pleas'd to ad‑
drefs to me, *De Itinere fuo Suecico*, in
octavo, in the Year 1720. That celebrated
Divine Dr. *a Melle*, of the fame City,
whofe Writings in Divinity, Hiftory, and
Antiquity, have raifed him into fo high
Efteem in the whole learned World, was
induced by the Perufal of my *N. H. E.*
to turn his Thoughts to the Study of *Fof‑
fils*. The firft Fruit of thofe Studies
was his *Epiftola de Echinitis Wagricis*,
4°. *Lub.* 1718. which he did me the
Honour to addrefs to me: As did
likewife the ingenious and curious
Mr. *Linckus* of *Lipfick* his *Epiftola
de fceleto Crocodili in Lapide* 4to. *Lipf.*
1718. The Count *de Schouberg*, Lord
Chamberlain to King *Auguftus*, and
Superintendant of the Mines in *Saxony*,
the richeft and greateft in all *Germany*,
fent me Samples of the Minerals and Ores
there, with their proper Names, and
thofe by which they are known to the
Miners; whereby I was enabled rightly
to underftand the Writings of *Kentman*,
Agricola, and others the moft learned,
accurate, and experience'd *Mineralifts* of
thefe

thefe laft Ages. *England* and *France* being ingag'd in a War, when firft my my Book came forth; and all amicable Communication betwixt the two Nations fufpended, 'twas not known there till the War was at an End. But, after that, it fell under the Cognizance of the Natu-ralifts of *France*, from whom I have fince receiv'd many Civilities: And in particular, from that great Mæcenas, *M. L. Abbe Bignon,* and fome other learned Ecclefiafticks, particularly of the Order of the *Jefuits* there; from Dr. *Juffieu* the King's Botanick Profeffor, who oblig'd me with Samples of fe-veral *French* Foffils, with very intelli-gent Accounts of them: But there being no confiderable Mines in that Country, the Curious there have not had much Op-portunity of carrying on thefe Studies. Monfieur *Valkeneir*, refiding for fome Years at *Zurich* in Quality of *Envoy* of the *States-General,* having perufed the *N. H. E.* and approv'd the Defign, pro-moted it with great Conftancy and Dili-gence, not only over the Country where he refided, but the greateft Part of *Eu-rope,*

rope ; and my Collection has been much enrich'd out of his Store. But the greatest and most beautiful Addition that ever was made to it, is owing to Signior *Agostino Scilla,* from *Rome,* he sending me not only all those noble *Fossils,* that he collected in *Sicily,* and publish'd in his *Lettera circa i corpi marini, petrificati* printed at *Naples* in 4to, 1670, but likewise the *original Drawings* of each, done by his own masterly Hand. I might mention to you several others ; but these will be sufficient to keep you in Countenance, and shew you that some, of the greatest Character in the whole learned World, have not disdain'd to embark in the same Bottom that you have done. As to those who have honour'd me and my Undertaking with their Patronage here in *England,* 'twould be too great a Task to recount all ; and therefore I must not mention any ; which will be the less Loss to you, as you are wholly a Stranger to them.

I am, &c.

Q   NUM-

NUMBER VI.

To the fame.

Of the Origin, Nature, and Conſtitution of the Belemnites.

S I R,

A S to what you ſay of Mr. *Lhwyd,* he was much prejudic'd, and rea-dy to catch at any Thing that might leſ-ſen the Authority of what I have deli-ver'd. I rank the *Belemnites* amongſt the *native Foſſils,* he would fain have it be thought to belong to the *Extraneous :* and his Book coming into every Body's Hands, ſome fell into his Notion ; parti-cularly Mr. *Butners* [a], who examines little, and is very ready to fall in with any thing that comes in his Way. That is far from your Caſe ; and tho' I have little Regard to them, I am ſo ambitious

of

[a] *De Corall. Foſſil. c. 6.*

Fig. 7.

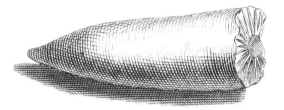

Part 2.

of the Opinion of a Man of your Weight, that I cannot contain my felf, from confidering what you write of the Subject. I grant indeed, as you obferve, that *my Hypothefis is not at all concern'd of which Side foever the Queftion is determin'd, and will not in the leaft be affected, tho' the Belemnites be not, as I have afferted, a meer Stone.* But I am *concern'd* for the *Truth,* and have that Regard to you, that I would have you think I did not affert that, without fufficient Grounds; nor has any thing hitherto been offer'd, that invalidates my Affertion. Whenever any Thing does, you will find me very eafy and ready to give it up.

Mr. *Lhwyd* [b] at fometimes fancies the *Belemnites* to be form'd in the *Pennecillæ Marinæ,* at others, in the Shells, call'd *Dentalia.* Thofe are Bodies as different each from other as well can be, and both differ fo much in Shape from the *Belemnites,* as to give little Umbrage

Q 2   to

---

[b] *Lythophil. Britan. p.* 115. 121.]

to the Notion that it could be form'd in
either. Befides, the Manner in which
we commonly find the Belemnites, fhews
plainly it was not form'd in any Shell.
When the Bodies fo form'd are found
lodg'd in the Strata of the Stone, tho'
the Shells wherein they were originally
form'd, be perifh'd and gone, the Stones,
moulded in them, are conftantly furroun-
ded with a Cavity, or a Space wherein
the Shell lay, which Cavity ever anfwers
to the Shape, and is commenfurate to the
Bulk of the Shell fo perifh'd. Now the
Belemnitæ are ever found contiguous to
the Mafs of the Stone, without any fuch
Cavity furrounding them. In this we
have Evidence of Senfe, and ocular De-
monftration, that the Belemnitæ were
not form'd in a Shell, or any external
Mould. Had they had any fuch Mould,
the Veftigia of it would have been eafily
enough difcern'd, and the Cavity where
the Shell lay prefently difcover'd. It
muft have been in fome very large Shells.
I have feen Belemnitæ near two Feet
long, and above two Inches Diameter in
the thickeft Part. Shells in which fuch

Bodies

Bodies could have been caſt, or the Cavi-
ties wherein they lay, would be ſo big,
thick and long, as to be deſcry'd without
Difficulty.

I am the more forward to think, that
the Reaſons upon which you found your
Suſpicions, are not very firm and clear,
becauſe you are ſo unſteady in your Opinion. You formerly thought the Belemnites a Horn ; now you fancy it a Tooth
of ſome ſtrange Fiſh, Bodies quite different in all Reſpects from each other.
Your firſt Arguments for its having been
an Horn, is, that it is in *Form of a Horn,*
whereas indeed there are different Species
of the Belemnites : And they differ very
much in *Form* from each other. The
three principal Species are, the *Conoid Belemnites,* which is the moſt common.
The *Belemnites fuſi-formis, J. Bauhini, de Fonte Bollenſi,* 4ᵗᵒ, and the *Belemnites Cylindricus in apicem utrinq;
deſinens.* If thoſe be all in *Form* of
Horns, every Thing is in *Form* of a
Horn. Your ſecond Argument is, That
'tis *lodg'd amongſt Shells, Teeth, and
other*

*other Animal Remains, found at Land,
and in the Strata.* In cafe this prove
them Horns, it will prove Pyritæ, and
many other mineral and metallic Bodies
to be *Horns*, or *Animal Remains.* For
thefe are found lodg'd amongft Shells,
Teeth, and other animal Remains, full as
frequently as the *Belemnitæ* are. Your
third Argument is, That the Belemnites
has a *horny Smell.* Now, if this be ad-
mitted, 'twill bring almoft half the native Foffils into the Clafs of Horns, that
Smell being common to *Stones*, and many
other *Native Foffils*, that have in them
Sulphurous or Bituminous Principles ;
and thefe they exert in greater Plenty, if
rubb'd and heated. Indeed Stone, when
firft taken out of the Earth, emits very
different Smells, *Ol. Worm.* mentions
[c] fome that emitted a Smell like that of a
Hog, which he therefore calls *Saxum
Suillum* ; other Stone, with the Smell of
Violets, *Lapis Violaceo Odore.* They
that are converfant with fubterranean
Things know, that not only Stones of
                                    various,

Mufcum Worm.

various, but even of the very ſame Sort,
emit very different Smells; ſo that no
certain Judgment can be form'd from the
Smell. Beſides, I muſt acquaint you,
that the Belemnitæ of *England* have
rarely any Smell at all. They are found
in great Numbers in Chalk, and I never
could perceive a Smell in any of theſe.
Thoſe that you found attended with that
Smell, had lain amongſt Saline, Sulphurous or Bituminous Matter, that had imparted it to them. But what ſeems to me
finally to determine this Controverſy, and
evince that the *Belemnites* is not a *Horn,*
is, that *Horns* are very ſeldom found in
the Earth. I have aſſign'd a plain Reaſon for that in the *Diſſertation prelimi-
nary to my Natural Hiſtory of the Earth,*
I have ſhewn there, that *Horns, Hoofs,
Teeth, Bones,* and other like *Animal
Subſtances,* being lighter than the common Sea-Shells ſubſided laſt, and conſequently being *lodg'd near the Surface* of
the *Earth,* and there expoſed to the
Weather, and external Injuries, are generally periſhed and deſtroyed; few of them
remaining at this Day. Whereas the Belemnitæ

lemnitæ are frequent, obvious, and occur almoſt every where. Nay, they are found to very *conſiderable Depths* in the *Earth,* which is owing to their ſpecifick *Gravity,* much greater than that of *Horns* or *Teeth,* but equal to that of *Talky Bodies,* in which Claſs I have rang'd them. *That* their *greater ſpecifick Gravity* furniſhes us with another Proof, that they are not *Horns,* or *Teeth.* A further Argument of which is, That they differ greatly in *Texture, Conſtitution and Subſtance* from *Horns, Teeth,* or any other like Parts of Animals. But they agree very nearly with ſeveral Minerals. I have ſeen ſome that are Semi-diaphanous, yellowiſh, and ſomewhat reſembling common Amber; which the Antients obſerving gave to both *Amber,* and the *Belemnites,* the ſame Name, *Lapis Lyncurius*; this Name importing that both were of an Hue yellowiſh, and like that of Amber; as are likewiſe ſeveral *Talcs, Spars,* and other Products of the Mineral Kingdom. Then, as to the *Conſtitution* of the *Belemnites,* if it be broke in any Manner, it is not tenacious and tough, as all Ani-

I   mal

mal Subſtances are : but *friable* and *brittle*, like *Talky* and ſuch other Foſſils. The Subſtance of it appears to be mineral even to the View ; and this is confirm'd both by the Operation of chymical Menſtrua, and every other Teſt. Its Texture is directly contrary to that of Teeth, and other hard Animal Subſtances, ſtriated acroſs ; the Fibres diametrically interſecting the Axis of the Body ; whereas the main Fibres of Teeth, Bones, Horns, Hoofs, Claws, Nails, and all hard Animal Subſtances run the quite contrary Way, and parallel to their Axis. But the ſtriated or *fibrous Talc*, the *Gypſum Striatum, talky plated Spars,* the *Aſbeſtos, Alumen plumoſum,* the *ſepta* of the *Ludus Helmontii,* the *Pipes* of the *Lapis Syringoides,* the Cruſts of the *Hæmatites,* and ſeveral other Talky Minerals, that might be recited, have their Fibres running in a tranſverſe Manner, like thoſe of the *Belemnites.* A remarkable Inſtance of this Texture we have likewiſe in ſome *Stalactitæ,* conſiſting of a *Talky Spar,* and found hanging down from the Tops of Grottos under Ground. There

R   are

are, in my Collection, feveral that are
ſtriated acroſs.

THESE Things rightly conſider'd, I
hope I ſhall not be accus'd of Lightneſs,
or Precipitation, in judging the *Belemnites* allyed to the Foſſils of *Talky Conſtitution,* as I have done. It has nearly
the ſame *ſpecifick Gravity* that the
Talky Bodies have, and is much heavier
than Horns or *Teeth.* Then 'tis exactly
of the ſame *Nature, Texture, and Conſtitution* that they are, and different in all
thoſe Reſpects from Horns, Bones, or
Teeth. Nay, I am perſuaded, the Arguments that I have offer'd, taken from
the Shape of the *Belemnitæ,* are ſufficient to ſatisfy any reaſonable Man, and
indeed amount to near a Demonſtration,
that they cannot have anſwer'd the Ends,
nor ſerv'd for the Uſes of *Teeth,* either
to ſeize the Prey, or to chew it.

BUT, tho' it be certain, that the Belemnitæ have now none of them, any
Thing of *Animal Subſtance* remaining,
I allow it does not thence neceſſarily follow,

I

low, that they may not have been of
Animal Origin; but 'tis very ſtrange,
they ſhould *all of them* be thus chang'd;
of which we have no Inſtance in any
other Body whatſoever. 'Tis indeed not
uncommon to find Shells of various
Kinds quite chang'd, the Teſtaceous Subſtance diſſolv'd, and a Mineral Subſtance
ſubſtituted and repoſited in the Room
and Place of it. Nay, there are digg'd
up, Parts of Trees ſo chang'd; and I have
ſhewn ^d how theſe Changes were brought
about. But then, the Inſtances of theſe
are are very few in Compariſon of the
whole : And for one Shell that is thus
chang'd, there are found hundreds that
are not chang'd at all. Whereas the Belemnitæ are all changed, if any of them
are.

WHAT I here offer, I intend as a Pre-
lude and Introduction to what I am about
to deliver in Anſwer to the Argument
urged from the *Tubuli Vermiculares,*
and ſmall *Oyſter-Shells* that are ſome-
R 2   times

---

^d *Anſwer to Camerar. Part* 1. §. 6.

times found adhering *externally* to the *Belemnites*. For, from this *Phænomenon*, some have haftily infer'd they are of marine Origin, and that thefe Shells were affix'd to them in the Sea before the Deluge. That will not by any Means follow from this; fince there are *Flints, Pyritæ,* and other *native Foffils,* that were never exifting in thofe Seas, that yet have Sea Shells adhering to the *Outfides* of them: And fuch I have in my Collection. For thefe, being ftony and Mineral Nodules, among which I have ranked the Belemnites, were form'd during the Time that the Water was out upon the Earth [e] : And the Matter which conftitutes them, then concreted, and affix'd to thefe Shells.

But there may be a Teft fettled, whereby this Affair may be fully determin'd, and it may be afcertain'd, whether the Shell upon it, or the Belemnites, was form'd firft. The Shells that affix themfelves unto, and grow upon Rocks, Stumps

[e] *Nat. Hift. Earth, Part* iv.

Stumps of Shrubs, and other fix'd Bodies, upon the Sea Shores, conform themſelves in their Growth, ſo exactly to the Surface of the Body on which they grow, as to take the Form of it. Now, if thoſe on the Surface of the *Belemnites* have done the ſame, and exhibit conſtantly the Lineaments of its Surface, then they were form'd ſince the Belemnites. But if this, in thoſe Parts where it is contiguous to the Shell, be not, as it uſually, and naturally is, ſmooth and plain, but exhibits the Lineaments, or any Impreſſion of the Shell, then 'tis certain the Belemnites muſt have been form'd ſince the Shell: And much more, if there be Shells found included in the Subſtance, or incorporated with the Maſs of the Belemnites. As, in all my Studies and Searches, I have nothing but the Truth in View, I willingly ſubmit to this Teſt, for the Deciſion of this Affair, and to further Inquiry. For I have ſo ſeldom found *Belemnitæ* with *Shells* upon them, that I have not Obſervations enow of my own to determin it. There is but one in my Collection, that hath a few very ſmall
Shells

Shells upon it; and I am unwilling to break it to make the Experiment.

THAT you may fee I have not been without Thoughts of this Subject, near twenty Years ago, when I was drawing up my *Catalogue of the Foffils of England,* taking notice of thefe Shells affix'd to the Belemnites, I enter'd there a Suf-fpicion grounded on this; with a Note for further Inquiry, *Whether the Belemnitæ may not have been originally Horns, or other like Animal Appendages, of of which we fee, by the Afteriæ, Eutrochi, and many more, there are, or have been, vaft Numbers at the Bottom of the main Ocean, that never appear upon the Shores.* Nay, Sir! I will fling you in of Courtefy, another Note that I made at the fame Time, *(Viz.)* " The Belem-
" nitæ fometimes appear to have been
" comprefs'd, crack'd, and deftroy'd;
" which is what I do not remember ever
" once to have obferv'd in any Foffil that
" was not form'd in an Animal Mould.
" But, in thefe, in Flints. form'd in
" *Echini,* and fome others, there are fuch
" In-

" Inftances ; " of which there are Accounts in the *fecond Part* of that *Catalogue.*

I am fo ambitious and defirous, Sir! that you fhould have full Satisfaction, that I will proceed a little more particularly to examin the Notion, that the Belemnitæ have ferv'd as *Teeth.* Now, of the many Hundreds that I have feen of thefe Bodies digg'd up here, and brought from Abroad, I never faw one that had the leaft Appearance of a *Fang* or *Root,* whereby it might be fix'd and detain'd in a *Jaw.* Whereas the *Teeth* of all Creatures that I have obferv'd, as well thofe that are the Product of the Water, as of the Land, have all *Roots,* or fome Signs of their having been connected to a Jaw. I know it will be faid of the *Belemnitæ,* that the Roots are broke off, and loft. But tis ftrange, of fo great Numbers as we find, there fhould not be the leaft Sign, or Remain of a *Root* on any. The Cafe is different in all other *Teeth,* as of *Sharks,* and other *Fifhes,* and indeed all other Creatures digg'd up out of the Earth ;

Earth ; thefe being commonly found with the *Roots* on, or, at leaft, with fome *Remains* and *Signs* of *Roots*.

THEN, there is *one Kind* of *Belemnites* that is of fuch *Shape*, that I think it could not have ferv'd for a *Tooth*, or even poffibly have been fix'd in a Jaw; I mean the *Belemnites fufi-formis* of *J. Bauhini*[f]. This terminates in an *Apex,* or *Point,* at *one End*; which, if any, muft have been the Tip, or upper Extremity of the *Tooth*. But the Part of the Body, next this, is turn'd crafs and thick; and the other for at leaft half the Length of the Body, very flender and thin. Now, tho' this was the Root, of which yet there is not the leaft Appearance, it being of the very fame Conftitution with the reft of the Body, which the Roots of the Teeth of thefe Fifhes that I have feen, never are : Or tho' there was, at the Extremity of this, a Root annex'd, and fince broke off, the contrary of which may be demonftrated meerly from View of

[f] *De Fonte Bollenfi. Pag* —

of ſeveral in my Collection, they being
at this End ſo very ſmall, that there was
not Scope for Hold ſufficient to connect
or fix it to any Thing. I ſay, which
Way ſoever it be ſuppos'd, intire, or broken, that Moiety of this Body, which
muſt be imagin'd, if any, to have been
next the Jaw, is ſo ſlender and ſmall, that
it is demonſtrable it could be of no
Uſe, and that the leaſt Force would break
it; eſpecially if it be conſider'd of how
tender and brittle a Nature it is. Whereas the other Moiety of this Body is ſo
tumid, thick and groſs, that it could never be got to enter the Prey for taking of
it, which theſe, if Teeth, muſt have
ſerv'd for, without a very great Force,
and ſuch, as the other Moiety could never have Strength near ſufficient to
ſuſtain, without being ſurely broke in
the very firſt attempt. So that 'tis evident this Body could never have ſerv'd
as a Tooth.

THE third Sort of *Belemnites* is very
nearly of a *Cylindrick Form*, only terminating at each End in a *Point* or *Apex*,

S   very

very fhort, fo as rarely to exceed $\frac{1}{10}$ of an Inch of Length. This is of the fame Nature, Texture, and Conftitution with the two precedent Kinds: And this I have found frequently *intire* ; but I have never found any of the *Conic* that was fo, tho', as has been intimated, they occur frequently, and in great Numbers ; which I can hardly fay of any Sort of Foffil befides, either native or Extraneous. For which Reafon fome have fufpected that the common or Conic Belemnites is broke in two about the Middle ; and that it terminated originally, and while intire, like that above fpecified, in a Point, at both Ends. Be that as it will, the Species that I am now treating of , has not the leaft Appearance of its having had any Connexion with a Jaw. Nor indeed is a Body of fuch a Shape, by any Means capable, either of being fix'd in a Jaw, or of taking of Prey, or of chewing of it. Indeed the *common Belemnites* is not much more capable of anfwering either of thofe Ufes. It is generally fo blunt at the End, as not to be capable eafily to enter the Prey ; and yet
                                                    not

not blunt or flat enough to mafticate and chew it. Befides, both this, and the other two Kinds, are ever *ftreight*; whereas the *Dentes Apprehenfores*, of all the *Fifh of Prey* that I have ever feen, are, the better to fit them for taking the Prey, in fome degree crooked. I wifh my Deference to your Judgment, and the Zeal I have to give you full Satisfaction, have not drawn me on fo far, as, inftead of that, by this Time to have given you Pain, and tired out your Patience. *That* I muft leave to your Goodnefs; but, for fear of the worft, I will defer my Returns to what you are pleas'd to command in Relation to the *Coralloids,* or *Coralls digg'd up at Land,* to fome other Opportunity.

I am,

always with great Regard,

Sir, *&c.*

NUMBER VII.

*Of the Coralloids digg'd up at Land ;
the Nature and Origin of them.*

SIR,

WHILE there are fo many forward
to write, and think themfelves
qualify'd for that Purpofe, fo foon almoft
as they turn their Thoughts to any Subject, and without, firft, being at the
Trouble of duly apprifing themfelves of
it, or of what others have deliver'd concerning it, the Minds of their Readers
muft be in a perpetual Maze, and Truth
upon a lubricous and very unfteady Foun-
dation.   The more fo, as there are fome,
who, tho' really much better Judges than
the Authors they read, without Sufpici-
on, or due Examination, fall into their
Sentiments, and adopt their Notions.
'Twas this Way, Sir! that I am perfua-
ded you fell into yours, of the Origin of
the Coralloids, as one or two other very
excel-

excellent Men have done, I mean, on Perufal of Dr. *Buttners Corallographia Subterranea.* That Author follows me, tho' he be not pleas'd to refer to my Book, in moft other Things; and thinks, fo far as he is wont, he does fo even in this; failing on in a full Gale of Fancy, and judging of Things pell-mell, he ftumbles, thro' meer Inadvertency, into the Notion, that the *Foffil Coralloids* are of *Antediluvian Origin*; and were by the Deluge brought out of the *Sea*, along with the Shells, and other *Animal Subftances*, to Land. Indeed, that he may proceed in Form, and the more like an Author, he brings in what he is pleas'd to call *Arguments*, in Support of this. But, being on no very high Guard, unluckily they either prove nothing at all; or elfe the quite contrary of what he alledges them for: His firft method of arguing is from *Similitude*; and comparing the *Foffil Coralls* with the *Marine*. He avers both have what he calls a *Bafis* or *Root*. That is commonly true of the *Marine*; but of the *Foffil Coralls,* he gives not fo much as one fingle Example that is clear and plain;

plain; nor of the Multitudes that I have feen, have I ever met with one. His fecond Argument is, that *both* have been *once* foft, or in a State of *Solution*. That I have prov'd very fully; but it makes nothing for his Purpofe; the Queftion is not about the *Fact*, but the *Time*. No Body doubts, but that they were *foft* at the *Time* of their *Formation;* all Things in Nature whatever, are fo; but that *Time* was not, as he prefumes, *before*, but *during* the *Deluge*. He proceeds in his Way of comparing the *Marine* with the *Foffil Coralls*. Some of thofe, he fays, have a Tendency to a *Vegetable Form*, they have *Trunks, Knots, Branches;* fo likewife have thefe: Others of them have *Pores, Stars,* and other Accidents, wherein they agree with thefe. But then he knocks down all again, and comes over to me, when he avers, that the *Coralls found at Land*, are of a real *ftony Nature*, and *chiefly of Flint*. If this be fo, they are as different as well can be from thofe *found at Sea*. He never faw one of thofe of *Flint*. However that be, he is peremptory as to the
Foffil

Foſſil Coralls: And goes on to aſſert, That *Flint is nothing elſe but an Antediluvian Corall.* Cap. vi. §. 2. Now *Flint,* or *Chert* is found in Form of *Strata,* as well as in *Nodules* of all Forms, of which ſome few are jagg'd and uneven, which are what I ſuppoſe he calls *branch'd.* So that if Flints are Coralls brought out of the Sea, Free-Stone, Marble, and, to be ſhort, every Thing elſe that is either in Form of *Strata,* or *branch'd,* muſt, by this Way of Reaſoning, be brought thence too. His next Argument is, That the *Foſſil Coralls* are found lodg'd in the *Strata* along with *Shells,* and other *Productions of the Sea.* The Fact indeed is ſo; and it has been obſerv'd a thouſand Times, that there is in the Strata ſuch a Confuſion of Things of the moſt different Nature, and Origin, Animal, Vegetable and Mineral; but whoever made an Inference like this from it before? There are found lodg'd in the Strata with *Shells, Nodules* of all Kinds, *Stony, Mineral, Metallic*; does it follow that theſe were brought from Sea, becauſe the Shells were? If it do, all Bodies whatever

ever were brought out of the Sea. Dr.
Buttners rejects *Chymical Trials* of the
Coralloids, Cap. 6. §. 17. except in one
Cafe, which makes for his Purpofe,
where he alledges in Proof, that *Flints*
are *Coralls*, becaufe *they* will *calcine* as
well as *thefe*. Cap. vi. §. 8. which, that
I may note that by the by, brings
Stones, and all other Bodies that may be
calcin'd, into the Clafs of *Coralls*. In
fine, his only grand, and, as he thinks,
infallible Argument, is founded wholely
on their exterior *Form*, and *Structure,*
tantùm ex ftructurâ Coralliorum marino-
rum, tanquam notis characterifticis cer-
tiffimis noftra judicemus Foffilia. Cap. v.
§. xvii. He neither cares to admit *Chy-*
mical Trials, nor bring both to the Teft
of their *fpecific Gravity,* nor indeed any
other, whereby Judgment may be form'd,
of their *interior Conftitution, Subftance,*
and the *Matter* of which each are compos'd ; tho' that be the only fure Way to
fhew the *true Nature* and *Origin* of both.
To that therefore I fhall have Recourfe :
Of the Multitudes that I have obferv'd,
I never light of fo much as one fingle
Foffil

Foſſil Coralloid that agreed with the Marine, or was of the ſame *Subſtance* and *Conſtitution*. How greatly they differ from the Marine, and indeed from each other in *Subſtance,* may appear from the following Inſtances. Many of them conſiſts wholly of a *Sparry* Matter. Others of *Cryſtall*, ſometimes very near clear and pellucid. Some of them have their conſtituent Matter of Flint, others of Agat. Others of Vitriolic, and the like Salts, that ordinarily in Tract of Time moulder, liquate, and fall to Pieces, after the Manner of the Vitriolic, and other Salts in the common Pyritæ. I have ſeen *Foſſil Coralloids* that have been compos'd of various Sorts of Mineral and Metallic Matter, that yet have been form'd into Shape of the *Marine Mycetitæ, Aſtroitæ,* and other like *Coralls.* Now all theſe have been form'd out of the diſſolv'd Mineral, and Metallic Matter in the Water of the Deluge [a]. The Antediluvian Coralls were like all other ſolid ſtony Bodies then in Solution in that Water ;

T and

Nat Hiſt. Earth. Part. iv.

and might concrete again, and form
true Coralls, there, as well as in the Sea-
Water. Doubtlefs it did fo; but that
Matter was in fo fmall a Quantity, and
bore fo little a Proportion to the Mineral
and Metallic, with which it was then
mix d and confus'd, as now rarely, if
ever, to be met with. I never found one
Sample compos'd of it, pure and diftinct.
Which cannot be thought ftrange, if the
Antediluvian Coralls were all diffolv'd
and deftroy'd. Whereas, if they had
been preferv'd, and, as Dr. *Buttners* fuppofes, brought, along with the Shells, to
Land, they muft have been now found
commonly there, as well as the Shells.
They would be full as eafy, or indeed
more eafy to be difcover'd, than the
Shells; not only as they muft have been
in great *Numbers*, but very many of
them are of *Colours* that foonest ftrike the
Eye, and are the moft eafily difcern'd.
Such are the *Fiftularia purpurea* of *Ferrante Imperato*, and the *Red-Corall*; of
both which there are fuch vaft Quantities found in feveral *Seas*, particularly in
the *Mediterranean*, on the Coafts of
                                    Spain,

Spain, Italy, and *Sicily.* Then, many Kinds of the *Marine Coralls* are very *large,* fo that, had they been brought forth, and left at Land, they muft have been obvious, and very eafily found out. I need go no further for proof of this, than to the *Aſtroites Maritimus Coralloides undulatus major,* or, as 'tis commonly call'd, the *Brain-Stone.* This is found in great Numbers, in *feveral Seas:* And I have feen of all Sizes, to twice, nay, thrice the *Bigneſs* of an *Ox's Head.* Surely fuch Bodies as thefe, were there any, could not be hard to fpy out.

Upon the whole, I think 'tis very evident, that there are few, if any, *true Sea Coralls,* ever *found at Land.* Confequently thofe that we do actually find were not brought from Sea: And Dr. *Buttners* is led into his Error, by taking a meer *Cloud for Juno,* Bodies that had only fome flight *exterior Refemblance* of Coralls, but nothing of their *Subſtance* or *Conſtitution,* for real Coralls. When the *Spaniards* firſt took

Pof-

Poſſeſſion of *Mexico*, amongſt other Things new and ſurprizing, they found in the Gardens of the *Americans*, plac'd for Ornament, in a very elegant and beautiful Manner, *Artificial Flowers*, which they had made of *Gold*, ſo nearly approaching, in exterior Form and Shape, the true, as to cauſe much Admiration in the *Spaniards* ; as near indeed, or perhaps nearer than the Foſſil Coralloids do the Marine Coralls.   But yet I have not heard, that any of the *Spaniſh Philoſophers* fell into the Speculation that theſe fine *Gold Flowers* were brought forth of *Seeds*, as the natural were.   Tho' had theſe Gentlemen done ſo, they had full as much, and indeed the very ſame Reaſon of their Side, that Dr. *Buttners* had ; and he might juſtly have claim d the Honour of being added to this *Hiſpano-American Sect*.   As things now ſtand, I'm as much puzzled to find out, in what *Sect* of *Philoſophers* to *range* a Gentleman ſo anomalous as he is, in what Claſs of Foſſils to range the Belemnites.

I wiſh,

I wifh, Sir! that I have not, by this
Time over-convinc'd you, and brought
you to your *Ohe ! jam fatis eft, Ohe!*
Tho' it be fo, that I have, I ought to
make no Apology: You have put me
upon a Sort of Force. If the *Belemnites* fhould, tho' I fee no likelihood of
that, prove not to be a native Foffil, no
more is needful than to change its Rank.
You own your felf it affeĉts not the
Whole. But, as to the *Coralls*, in cafe
thofe *now digg'd up* be the *Antediluvian,*
they are a lafting and ftanding Monu-
ment and Evidence, that there inter-
ven'd no *Diffolution;* or, at leaft, that
it was not *univerfal.* For, if one Set of
Bodies, really ftony, could fo maintain
their *Solidity,* and fecure themfelves a-
gainft the common Law then in Force,
fo as to continue intire and *undiffolv'd,*
why might not any, or all the other
Setts do fo too? You muft not therefore
blame me. You fee the Queftion is of
the utmoft importance: And you have
made it neceffary for me to give you all
this Trouble to defend it, and fhew you,
that

that thefe *now digg'd up,* are *not real Coralls,* but of very *different Nature;* which I hope I have done to your Satif-faction, and fhall rejoice to hear that.

I am, Sir, *&c.*

NUMBER VIII.

Concerning Coralls, Corallin, and other like Bodies form'd at Sea.

PREFACE.

THO' thefe elegant, beautiful, and very extraordinary Bodies, have been much admired in all Ages; yet, lying far out of the Way, being hard to come at, and their Growth under Water, where accurate Obfervations cannot well and eafily be made, 'twill not be thought ftrange, that our Accounts of them are mighty defective, and that little Pro-grefs has been hitherto made in the Natu-ral Hiftory, and the Procefs us'd in the

Pro-

Production of them. 'Twas this which first drew my Thoughts to the Study and Consideration of them. But *when afterwards I found at Land, and in the Bowels of the* Earth, *various Bodies carrying some* exterior Resemblance *of the* Marine, *it ingag'd me to allow them some further Consideration, and carefully to compare both together. In order to this, the Coasts of* England *yielding very few Coralls, I had Recourse to my Friends in Foreign Parts, where these Bodies are found in greater Plenty and Variety; and by their generous Contributions, my Collection has been so far increas'd, as to exhibit Phænomena sufficient to point forth the* Process *of* Nature *in the* Formation *of these* Bodies. *Of this I have prefix'd some Account to the* Catalogue *of the* Coralloids *digg'd up in* England. *The following* Directions *were drawn up at the Request of Sir* Hen. Newton, *then* Envoy *of* Great Britain *in* Tuscany, *on the Coasts of which Country these Bodies are more frequent.*

DIRE-

DIRECTIONS

For making Obſervations on Co-
ralls, Corallin, and other like
Bodies.

I. **G**ET an Account of the ſeveral Places in which Corall is
found.

II. Also of the various Kinds of Corall found in each Place: Their various
Shapes and Colours.

III. And of the Manner and Poſture
in which the Corallin Bodies, particularly the Shrubs, grow; whether erect,
horizontal, or hanging down like Iceicles
from Jetts of Rocks.

IV. At what Depth the Corall grows.
And whether only in Parts of the Sea
that

that are under Shelter, and quiet, as in Creeks and Bays ; or in thofe that are more expofed and difturbed, as off Promontories, and the like : Or in both indifferently.

V. Of the Colour, Nature, and Conftitution of the Rocks and Cliffs, upon, or near which the Corall grows. Particularly obferve, whether there be any Red Stone, or other terreftrial Matter that is Red, near thofe Parts where the Red Corall grows.

VI. What is the Senfe and Opinion of the *Pefcadori,* or Corall-Fifhers, and of other more intelligent and curious Obfervers of the Growth and Formation of Corall ; of the Matter whereof it is formed : And of the Place from which that Matter is deriv'd.

VII. To what other Bodies is Corall found growing befides Rocks, loofe Stones, Pebles, Flints, and Shells.

VIII. Is

VIII. I s there any Way of making Judgment, whether the Corallin Bodies grow quickly or flowly : And in what Space of Time they are formed.

IX. Are the Corallin Bodies ever found broken and beat off the Rocks by the Agitation and Motion of the Sea in Storms.

X. Wнат are thofe Creatures that the Corall-Fifhers call Worms, that fcoop, bore, and hollow the Coralls.

Directions *for making Collections of Corall, Corallin, and other like Bodies.*

I. SEND Samples of Corallin Bodies of all Sorts, all Sizes, Shapes and Colours.

II. Also

II. Also of thofe which are various; feveral Sorts, or Corallin Bodies of feveral Colours, growing together.

III. And Samples fhewing the Manner of the Growth of the Corallin Bodies, upon Stones, Shells, or any other Things.

IV. Likewise of all Bodies whatever, that are drawn up by the Corall Fifhers; not only the Corallin Shrubs, Red, White and Black; but of the *Corallo Stellato, Articolato, Hippuris Saxea Pori, Millepora, Retepora, Frondepora, Madrepora, Tubularia,* [mention'd by *Ferrante Imperato, Hiſtoria Naturale.* L. 27.] *Fungi Marini,* [mention'd by *Padre Boccone* in his *Obſervation. Nat.* 12°. his *Muſeo di Fiſica,* 4°. and his *Recherches and Obſerv. Naturalles* 8°.] *Brain-Stones, Aſtroitæ,* and all others: And even of the Corallins, Sea Fans, Halcyoniums, Sponges, Moſſes, Algas, or Fucus's, Sea Shrubs, and Sea Weeds, of all Kinds: As alfo of the Shells, and

Stones

Stones of all Sorts.   In a Word, Samples
of all Bodies whatever, that are dragg'd
up in the Corall Fiſhings : And particularly of all thoſe Bodies that the *Peſca-*
dori call *Ravano.*

V. Send Samples of the Rocks in the
Neighbourhood of the Corall-Fiſhings ;
and of any other terreſtrial Matter, out
of which the Corall may be imagined to
be formed.

THE
PREFACE.

THOSE who travel and pass suddenly from Place to Place, have less Opportunity of informing themselves of all Circumstances of Things, than they that dwell, and are constantly upon the Spot. For which Reason, wherever I found, that either the Proprietor himself, some other Gentleman that happen'd to live near, or the Steward and Overseer of the Mines, had Curiosity and good Disposition, I engag'd them to make Observations and Collections; leaving with them Directions *for the Purpose. By this Means I receiv'd some Additions; but not to near the Number that, were Gentlemen better appriz'd of the Uses of these Things than they commonly are, I might have reasonably expected.*

Partly

The PREFACE.

Partly therefore on this Account, and partly because my Affairs call'd me up to London, *before I had compleated what I first design'd, and visited all the Mines that I intended, I concluded to send Persons on Purpose to all Parts where I wanted further Satisfaction and Intelligence; which I did at my own private Expense.*

Of these, the first that I imployed thus was Mr. Thomas Lower, *my Servant, a young Man, related to* Dr. Lower, *and of a good Family in* Cornwall. *Thither I dispatch'd him with* Directions *to make Observations in the Tin Mines, and to collect all the Ores and other Minerals he could procure. Being a sensible Man, and very careful, he executed his Commission with that Success, that he not only made for me a large* Collection *of Samples, well chosen, but a great Number of pertinent* Observations *of the Water in the Mines, and the Condition of Things there very much to my Purpose.*

This encourag'd me to proceed, and send others on the same Design; which I did,

The PREFACE.

did, but none to better Purpose than Mr.
John Groom, *and* Mr. Richard Meales,
*two learned and ingenious Gentlemen,
who were pleas'd to travel over a great
Part of* England *for me, and particularly all the northern Countries.*

'*Twas for the Service and Conduct of
these Gentlemen, that the following Directions and Queries were drawn up. Those
relating to the Oeconomy of the great A-
byss, Steams and Damps in Mines, Fogs
and Mists on great Mountains, and to
Meteors, were added by Command of the
Lord Bishop of* Man, *and Sir* George
Wheler, *two Persons not more illustrious
for their Piety, Virtue, and Knowledge
in their own Profession, Theology, than
their Insight into all other good and useful Learning. Residing in Parts where
they had great Opportunities of making
these Observations, they were the more
capable of promoting my Design; and indeed I am oblig'd in Gratitude to acknow-
ledge they were two of the most generous
Benefactors to it.*

Brief DIRECTIONS

For making

Obfervations and Collections,

AND

For compofing a travelling Regifter of all
Sorts of Fossils.

I. *Of keeping a Regifter of the Foffils
as they are Collected.*

Y Means of Pafte, Starch, or
fome fit Gum ought to be fix'd
on each Sample collected, a bit
of Paper, with a *Number upon*
it, beginning with Nº. 1. and proceeding
to 2, 3, and fo on, in a continual arithme-
tical Series. Then, in the Regifter, enter *Numbers,* anfwering thofe fix'd on
the

Foffils, and under each Note, 1°. *what Sort* of Foffil or Mineral 'tis reputed to be. 2. Where 'twas found. 3. Whether there were more of the fame, and in what *Number* or *Quantity*. 4. Whether it was found on the *Surface* of the Earth: 5. Or, if it lay deeper, note at what *Depth.* 6. In what *Pofture* or *Manner* it lay. 7. *Amongft what* Sort of terreftrial Matter 'twas lodged: 8. Whether in a Stratum, or perpendicular Fiffure.

II. *Of Searches upon the Surface of the Earth.*

WHERE the Ground is covered with a *Turf* and *Herbage,* few Minerals are to be met with ordinarily, unlefs in fuch places as have been formerly ploughed. But where the Earth is difturbed and turned up by *plowing, digging,* or any other Means, there Minerals are frequently brought forth, and expofed to Light; fo that *ploughed Fields* ought not to be neglected; efpecially thofe that lye *high,* and are *raifed* above the neighbouring Plains and Valleys; for in fuch the loofe Mould

is

is waſh'd off by Rains, born down, and by that Means ſuch obſervable Foſſils of all Kinds, as lay within, near the Surface, are laid bare and uncovered.

But the Tops and Sides of *Hills* and *Rocks*, the Earth and Sand being perpetually worn and beaten down by Showers and Storms, never fail of a *more plentiful Shew* of theſe Bodies, and a fuller Gratification of the Curioſity of an Enquirer.

Then, for the *Shores* of *Rivers*, and of the *Sea*, and the *Cliffs* adjacent, theſe uſually afford Variety of Minerals, and other obſervable Bodies; the Water waſhing and bearing off the Earth in which they were originally lodged, by that Means expoſing them to View. 'Tis here we find great Numbers of Pyritæ, and other Mineral Nodules: Nay, oftentimes Jett, Amber, Agates, and Stones of much greater Worth: As alſo Shells, Teeth, and other like Things that came firſt forth of the Cliffs and neighbouring Earth

Earth in which they had lain ever fince the Deluge.

But the Bowels, and *deeper Parts* of the Earth, contain the greateft Number and Variety of thefe Bodies. And, for Difcovery of them, Recourfe muft be had to fuch Places where there is finking for *Metalls, Marble, Stone, Alabafter, Coal, Gravel, Chalk, Oker, Fullers-Earth, Clay* for Pots, Tyles or Bricks, *Sand, Marle,* or the like: Or when there are Wells making: And in fhort, wherever there is *Digging* upon any Occafion whatever.

III. *The Method of making* Obfervations *in* Mines, Pits, *and* Quarries, *and of compiling a* Regifter *of them.*

I. The firft Thing to be taken notice of is the Place and Site of the Mine, Pit, or Quarry, whether it be in a *Valley*, on a *Plain*, or on an *Hill.*

II. Whether the *Defcent* into it be perpendicular, and by a downright *Shaft,* or or the Paſſage down be only upon a ſhelving or inclining Way.

III. Note the *Extent* of the *Aperture* of the Quarry Pit or Mine ; as alſo of the ſeveral *Vaults* in it, and how far the Strata or Beds of Stone, Earth, *&c.* are *extended* and expoſed to View in Front.

IV. Then proceed to conſider the ſeveral Strata, remarking, 1°. how they *terminate,* or whether they be diſtinguiſhed from each other only by the *different Nature,* Colour and Conſiſtence of the Matter that conſtitutes them ; or are ſevered by Joints, Partings, or *Fiſſures* running betwixt them. 2°. In what *Poſture* the Strata lye, whether level and horizontal, or *inclining.* If the later, note to what Point of the Compaſs the Dipping or *Inclination* bears, and how many Inches or Feet the Stratum ſinks below the Level, or horizontal Plane, in ſome certain determinate Space ; as ſuppoſe in the Courſe of 4, 6, or 8 Yards. 3°. Whether all the *Strata* lye *parallel,* and conformable to *each other,* and to the exterior *Surface* of

X the

the *Earth.* 4°. After this, come to a
clofer Examination of each fingle Stratum
apart, beginning at the *Top*, and taking
them one by one in Order as they lye,
quite down to the *Bottom* of all; noting
every Particular obfervable Circumftance
in each; e. gr. the *Thicknefs* of the Stratum, and whether it be of equal Thicknefs
in all Parts of it. The Thicknefs of the
feveral Strata, added together at laft, give
the *whole Depth* of the Mine, Pit, or
Quarry. Or if thefe cannot conveniently
be meafured fingly, the whole Depth may
be taken; the *Confiftence* and Conftitution
of the Matter or Bodies that compofe the
Stratum, 1. Whether it be loofe and foft,
or hard and folid; or partly loofe and
partly folid; what particular *Sorts of
Matter* each Stratum is compofed of, e. gr.
Marble, Alabafter, Free-Stone, Lime-
Stone, or any other Sort of Stone; or Coal,
Ochre, Chalk, Sand, Gravel, Clay, Marle,
or of Metallick or Mineral Matter : What
other *Sorts of Foffils* are embodyed in, or
lodged amongft the ordinary Matter of
the Stratum; e. gr. *Stones of an obfervable Figure,* as the Belemnites, Selenites,
Myce-

Mycetites, Corallites, Aſtroites, &c. or
any *Mineral Nodules,* ſuch as the Pyrites, Marcaſite, Hæmatites, Manganeſe,
Jett, Amber, Agate, Cornelian, Flints,
Pebles; or *Metallick Nodules,* or Lumps,
yielding Copper, Iron, Tin, Lead; or
any *Metallick or Mineral Matter* inter-
ſperſed in *ſmall Parts,* and mixed with
the Sand, Stone, Earth, or other common Matter of the Stratum; examining
whether the ſaid Metallick and Mineral
Nodules, or Matter, be chiefly of *one
Sort;* or, if of *ſeveral,* what *Proportion*
there is of each. Obſerve whether there
be any *Trees,* Nuts, Acorns, Fir-Cones,
Leaves, or other *vegetable Bodies* lying in
the ſaid Strata; or any *Teeth, Bones,
Horns, Hoofs,* or other *Parts of Animals*
of any Sort; or *Shells* of the *Cruſtaceous
Kind,* e. gr. Crabs, Lobſters, &c. or of the
teſtaceous, ſuch as Oyſters, Muſcles, Scal-
lops, Lympets, Perewinkles, or any o-
thers whatever. But with more particular Care examin the *Strata that lye deepeſt,*
and at or near the *Bottoms* of Pits, Mines,
and Quarries, to diſcover whether they
contain any *Shells,* or other like extra-

neous

neous Bodies. If the faid deeper Strata
be of *Stone* or *Marble,* break off Pieces,
and *wet* them with *Water*, to wafh off
the Duft or Powder that may cover and
obfcure them; then viewing them with
great Application, obferve whether broken Edges or other Parts of Shells do
not appear. The Shells, and other extraneous Bodies immers'd in Stone,
have oftentimes their Pores fo fatu-
rated with the fame Sort of Sand with
that which conftitutes the Stone; nay,
even their Surfaces are fo ting'd, and frequently fo much of the fame Complexion
with the Stone, that they are not to be
difcovered without a very nice and careful Examination. Laftly, Note in what
Numbers the faid *Vegetable and Animal
Bodies* are found; in what *Pofture* they
lye, and particularly, whether the *flatter*
and broader *Shells* (as likewife the flat
and comprefs'd *Mineral Bodies* of all
Sorts) be not repofited in fuch Manner,
that their *Flatts* are parallel, and conformable to the Surface of the Stratum in
which they are enclofed. Enquire whether there are latent in the faid Strata,
any

any Flints, Pebles, or other Bodies that reſemble, or have Marks or Impreſſions of Shells, or of Leaves, Teeth, Bones, &c. upon them.

V. Obſerve whether each ſingle Stratum of Stone, Marble, or other ſolid Matter be *whole* and *continuous,* or *broken* and divided by Clefts or *Fiſſures.* In caſe they are, take notice, 1°. whether the ſaid Fiſſures ſever, and paſs down thorow only *one,* or *more,* or all the Strata. 2°. Whether they be *perpendicular,* and tend upwards directly towards the Surface of the Earth, or Slant and decline. 3°. Of what *Wideneſs* and Capacity they are. 4°. In what *Number,* and at what *Diſtance* from each other. 5°. Whether the Strata, on one Side the Fiſſure, *an-ſwer,* tally, and fit thoſe on the oppoſite Side of it.

VI. Examin whether thoſe *Fiſſures* be *empty,* or contain any *Matter* in them. If the later, obſerve, 1°. of *what Sort* it is, whether ſome Kind of *Ore,* e. gr. of Lead, Tin, Iron, Copper, &c. or ſome

Mi-

Mineral, as Sulphur, Mundick, Marcafite, Calamin, Amianthus, Cobalt, Load-
Stone, Cinnabar, Antimony, Bifmuth,
Speltre, or Talc, Spar, Cryftall, or whether
there be Amethyfts, Topazes, Saphires,
Emeralds, *&c.* or common Salt, Nitre, Vitriol, Alum, *&c.* 2°. In what *Number* or
Quantity the faid Ores, Minerals, and
other Things are found. 3° In what *Order* they are repofited in the Fiffures.
4°. In what *Manner* and *Form* they appear; whether they lye only in *rude Maffes,* or are difpofed and fhot into any *obfervable Figures,* e. gr. Rhombs, Cubes,
Pyramids, *&c.* Whether the *Native
Metals* be ever found in Threads or
Plates or Maffes, fo *pure* and free from Admixture of other Matter, as to be flexible
or *malleable.* And whether any Part of
the Metallick or Mineral Matter be formed into *Stalactitæ,* or Bodies refembling *Icicles* hanging down from the
Jetts of the Fiffures, or vaulted Tops of
Grottoes ; or cover and *cruft over* the
Stone at the Bottoms and Sides of
them.

VII. Ob-

VII. Observe in *what Manner Water* iſſues into Pits, Mines, and Quarries; in what Quantity it enters; at what Time it is moſt plentiful; whether it be *pure* and taſtelefs, or *tinctured* with Salt, Nitre, Alum, Vitriol, or ſome Kind of Mineral.

VIII. Enquire, whether any Guſts of *Wind* be ever obſerved in the ſaid Pits or Mines, or any Sorts of *Damps,* or *Steams*; what are the *Signs* or preſages, and what the *Conſequences* and *Effects* of them; at what *Seaſons,* and in what Sort of *Weather* they are chiefly obſerved; what *Temperature* the Air bears, as to *Heat and Cold,* in Pits or Mines; and whether it be conſtant or changeable; in caſe the later, Information ſhould be got at what *Seaſons,* and upon what *Occaſions* thoſe Changes happen; as alſo, whether there ever be obſerved any *Steams, Damps,* unuſual *Heat* or *Cold,* or any other remarkable Accident in the *Bottoms* of Mines, Pits, or Quarries, a little before, or during the Time of *Rain,* Hail,

Hail, Wind, Storms, Thunder, or other
extraordinary *Weather* in the Air above.
[See the Appendix infra.]

THIS is the fitteft Conduct and Pro-
cedure I can pitch upon for their Obfervations and Enquiries; and what Intelligence and Information is gain'd by them
may be enter'd into the *Regifter* in the very
Method it muft needs arife by the Regula-
tion of the Courfe of the Obfervations
according to the foregoing Directions,
or as near as conveniently may be.
To which Purpofe that Regifter ought
always to be *ready* at Hand on thefe Occafions; and the Obfervations entred
upon the *Place*, for fear of Miftakes, or
Failure of Memory. At leaft, *Notes*
ought to be taken upon the Spot, and
they to be entered into the Regifter as
foon as may be, and while all is frefh in
Mind. In the tranfacting of this whole
Matter, great *Truth* and *Faithfulnefs,*
as well as *Exactnefs* and *Care,* ought to
be ufed; a Failure in either, tho' very
fmall, leading oftentimes into confiderable
Errors.

I   THE

THE Inſtances here pointed forth, and the Phænomena to be obſerved, are very *numerous* ; and 'tis not to be expected, that near ſo many can ever occur in *any one Pit* or Mine. Or if they do, there are few Perſons that have the *Leiſure*, or perhaps the *Curioſity* to attend to all of them. In which Caſe 'tis only deſired that thoſe *Inſtances* that do occur in any Place, whether they be more or fewer, be noted ; and ſuch *obſervable Bodies,* as appear, be collected. And for thoſe who cannot beſtow much, may at leaſt employ ſome Time in theſe Searches ; which, if they do, and are but Maſters of *Judgment* and *Thought* enough to make the Uſe of them, that they may eaſily do, they can never have Cauſe to think that Time miſpent. For *theſe Inquiries* tend not only to the promoting of *ſecular Profit* and *Advantage :* But, which is not leſs conſiderable, carry the *Mind* of Man into a Field of *Knowledge* that is extenſive, entertaining, and inſtructive, hardly to be expreſs'd by Words.

Y   BUT

But there are some of the Observations that cannot well be made by any but the very *Persons employ'd* in *Digging* and *Mining*. The *Adits* and *Shafts* of Mines are usually fenced and *covered* with Timber to secure the Earth from falling in; so that the Strata of those Shafts by that Means being concealed and screen'd from View, an Account of them can be had only from the *Miners*, and those who sunk them. But then the Strata at the *Bottom* of those Mines lye fair and open for Observation, and may be viewed oftentimes to a very great *Extent*. Again, there are other Things that require *Time*, and some considerable *Abode* in the Mines, Pits, or Quarries, to come to due Knowledge and Information of them. Such are *Winds, Damps,* and *Exhalations* in the Bowels of the Earth; the *Vicissitudes* and *Seasons* of them; the various *Temperature*, Heat and Cold of the Air underground, at *different Times*, and the like. These must be learned of the *Miners* and *Workmen*; and they may be likewise conferr'd with

with about thoſe Things that are *more obvious*, and liable to Obſervation. But particular *Care* ought to be had not to conſult or take Relations from any but thoſe who appear to have been both *long converſant* in theſe *Affairs*, and likewiſe Perſons of *Sobriety*, Faithfulneſs and *Diſcretion*, to avoid the being miſled and impoſed upon either by *Falſhood*, or the Ignorance, *Credulity*, and *Fancifulneſs*, that ſome of theſe People are but too obnoxious unto. And, after all, there ought to be a *Diſtinction* made in the Regiſter betwixt thoſe Obſervations *perſonally* made, and thoſe that are *communicated* by the *Miners*.

APPENDIX I.

Concerning Mines.

ENQUIRE of the Miners, 1°. Whether they have ever met with any Evidences of the *Growth* of any Metall or Mineral; and whether after a Stratum, Vein or Fiſſure, is once cleared, and the

Ore

Ore intirely taken forth, they ever after find any, either of the fame, or any other Sort, in that very Stratum or Vein.

II. What *Signs* in the Earth or Water, the Miners conduct themfelves by in their Working and Searches after latent Metalls and Minerals.

III. Whether there be any Thing particular and obfervable in the *Inftruments,* or in the *Methods* they make ufe of in *mining.*

IV. Whether they ufe any Sort of *Flux* in their *fmelting Works,* befides *Slagg* or *Cinders* ; or there be any Thing uncommon, and peculiar in the *Structure* and *Contrivance* of thofe *Works.*

V. Whether there be any particular and extraordinary mechanical *Inftruments* or *Artifices* made ufe of in their *Forges* or Furnaces.

VI. Whether the Perfons that frequent and work in the Mines be fenfibly
injured

iniured in their *Health,* by poifonous or
unwholfome Steams arifing thence : Or
the Air, Water, Herbs, or Fruits near
the Furnaces or Forges, be *noxious* or
offenfive to Men or Cattle.

APPENDIX. II.

Relating to fuch Fens, Boggs, *or*
Marfhes, *in which the* Peats *or*
Turffs *ufed for* Fuel *are got.*

1. OBSERVE their *Place* and *Site,*
whether in a Valley, on a Plain,
or an Hill.

2. THE *Bounds* and *Extent* of them;
and whether there be not Tracts of *Sand*
or *Earth* of a Nature different from that
of the *Turff-Earth* interpofed amidft it.

3. EXAMIN what is the *Thicknefs* of
the Stratum of the Turff-Earth ; and
whether it be of the fame Thicknefs in
all Parts of it.

4. WHAT

4. WHAT *Sort of Earth*, Sand, or other Matter lies at the *Bottom* underneath the Turff-Earth.

5. WHAT are the Properties, *Nature* and Conftitution of this Earth; and whether it be all of the fame, or of different *Sorts.*

6. ENQUIRE whether the Turff-Earth grow; or, what Evidences there are, that when it, or any Part of it, is cut and digg'd up, 'tis in Tract of Time repair'd and fupply'd afrefh.

7. WHAT *Springs*, or other Receptacles of Water there are in thefe Marfhes.

8. WHETHER there be any Bones, Teeth, Shells, or other *Animal Subftances* found lodged in this Earth; and at what *Depth*, in what *Manner*, and in what *Numbers* they are found.

9. WHE-

9. WHETHER any Trees, Shrubs,
Herbs; Fir-Cones, Nuts, Acorns, or any
other *vegetable Bodies.* Of what *Kinds*
they are, and whether there be of the
fame Kinds of Trees, Shrubs, *&c. now
growing* in or near thofe Marfhes; at
what *Depth* they are found, of what
Bignefs, and in what *Numbers.* In what
Pofture the Trees lye, in what Condition
they are found as to Firmnefs and Sound-
nefs: Whether the Roots be found yet
adhering to the Stump of the Tree, the
Trunk or Body being fever'd off from it:
Whether, if fo found, the Stump be in a
growing Pofture, ftanding up above the
Roots, or it he alfo fometimes *reverfed,*
and turned topfy turvy, with the *Stump
downwards,* and under the Roots: Whether *one Stump* with the Roots be not
fometimes found placed direftly *over
another,* or in fome *other Pofture,* wherein both could not naturally have grown:
Whether in any Marfh there be found
only the Roots, without the *Trunks* of
the Trees, or the Trunks *alone,* but *no
Roots:* Whether either Trunks, or Roots,
when

when firſt taken forth of the Earth, have any *Marks* or *Signs* of human *Workman-ſhip* upon them, appearing to be cut with an *Ax*, or *Saw,* or there be any *Cinders,* or other Evidence of Fire, evincing that Part of them hath been *burnt*.

10. By what *Means* do theſe Trees, and other Bodies ſeem to have been *re-poſited* in that Manner ; and what are the *Opinions* of the Perſons employ'd in digging the Peats, and of the near *Inhabitants* concerning them.

APPENDIX III.

Of Mountains, Rocks, and Cliffs.

1. OBSERVE the *Bigneſs* and *Height* of the Mountain; what Grottoes are in it ; what *Springs* ariſe upon it, and in what *Part* of it they are ; as alſo, what *Rivers* or running Waters have their Sources in it ; and what Quantity

tity of Water they difcharge Summer and Winter.

2. I f by Means of any *Fall of Earth* from it, the Mountain, Rock, or Cliff be laid *bare*, and its Strata expofed to View, or by the repeated *Battery* of *Rains,* or the Violence of the *Sea*, digging for Stone, Marble, or the like, obferve, 1. the *Pofture* of the Strata, whether *Horizontal, inclining,* or *erect*; alfo their *Thicknefs, Confiftence,* and *Fiffures*. 2. The feveral *Sorts* of *terreftrial Matter* of which each confifts, recounting them in the *Order* they lye. 3. What *Metallick* or *Mineral* Matter they contain. 4. What *Shells, Teeth,* or other *extraneous Bodies*.

3. Search carefully in all Places for *Shells*, and other *Marine Bodies* ; but more efpecially at and near the *Tops* and *higheft Parts* of Rocks and Mountains.

4. Enquire whether the *Tops* of the higher Mountains and Rocks do not emit *Vapours* in great Plenty, or there

Z   be

be not a *Cloud* hovering upon them before, or during the Time of *Rain, Hail,
Snow, Wind, Storms, Thunder,* or
other tempeſtuous *Weather :* Whether
from the Manner, Colour, Bigneſs, Duration of the Cloud or Vapours, any *Pre-
ſage* may be made *what ſort* of Weather,
e. gr. whether Wind or Rain will enſue;
or of *what Continuance* it will be ; whether the ſaid Cloud or Vapours appear
upon Change of Weather *conſtantly*, or
only at *ſome Times.* 'Twere much to be
deſir'd, that ſome *Perſon* living in View
of ſuch Mountains, would keep a *daily
Regiſter* of the *Weather*, and at the ſame
Time of all the *Phænomena* of the ſaid
Cloud or Vapour ; and if he be in View
of two or more ſuch Mountains, at once,
that he extend his Obſervations to all of
them.

5. Whether ever there be any extraordinary *Eruptions* or *Diſcharges* of
Water in conſiderable Quantity, out of
thoſe Mountains.

APPEN-

APPENDIX *to* *Page* 107 fu-
pra, *containing more full, expli-
cit and particular Inftructions
for making Obfervations concerning Fogs, Mifts, or Clouds,
feen frequently upon the Tops of
high Hills or Mountains.*

1. **O**BSERVE whether thefe Fogs
arife out of the Hill ; or whence
otherwife do the Vapours that conftitute
them proceed.

2. WHETHER they be feen hovering
over the Top of one only Hill, or of
more.

2. WHETHER the Fog on the feveral
Hills firft appear at the fame Time on
each, increafe in equal Proportion on
all, and decreafe likewife on each at the
fame Time.

Z 2   4. WHE-

4. WHETHER thefe Fogs be conftant Forerunners of Rain; fo that it never happens either in Summer or Winter, unlefs they appear before ; and whether Rain always follows whenever fuch Fogs appear.

5. OBSERVE how long they appear before the Rain falls.

6. WHETHER any Judgment can be made by View and Obfervation of thefe Fogs, of the Quantity or Duration of the Rain ; or whether it will be attended with Storms of Wind, or by Thunder and Lightning.

7. WHETHER the Rains that fall, feem to proceed from the Fog gradually diffufing it felf, and overfpreading the Country.

8. WHETHER the Barometer conftantly fall at fuch Time as the Fog rifes, and in Proportion to the Quantity of it, and rife

rife again at fuch Time as the Fog is difperfed and withdrawn.

APPENDIX *to Page* 101 fu-
pra, *containing more particular
Inftructions for making Obfer-
vations concerning Præfages of
Rain in deep Mines, great
Quarries or Coal-Pits.*

1. OBSERVE whether Wind, Rain,
Thunder or Lightning can be
foretold before they happen, by any Vapours, Steams, or Exhalations in the
Mines, Quarries, or Pits.

2. WHETHER it can be diftinguifhed
by the Manner, Colour, or Conftitution
of the Vapour that fhall enfue, whether
Rain, Wind, or Thunder.

3. WHETHER Judgment can be made
of the Quantity and Duration of the
Rain or Wind, by the Thicknefs of the
Vapour,

Vapour, the Continuance of it, or any other Way.

4. Whether the Vapour confifts fimply of Humidity ; or is alfo charged with metallick or mineral Steams.

5. Whether Rain conftantly enfues as often as thefe Vapours difcover themfelves in the Mines, and the Vapours conftantly forerun and appear before Rains.

6. Observe how long the Vapours difcover themfelves before the Rain falls.

7. Whether thefe Vapours be obferved only in fome, or in all Mines indifferently ; and whether they rife at the fame Time in all, fo far as Intelligence can be obtain'd.

8. Whether they are attended with any unufual Heat.

9. Observe wherein thefe Steams differ from thofe called Damps ; or whether

ther Damps greater or lefs, and Rain, conftantly attend each other.

10. Observe how the Barometer and Thermometer, as well thofe kept in the Mines, as thofe above Ground, are affected during the Afcent of the Steams and Damps, and during Rain; as alfo before and after.

NUM-

NUMBER IX.

An Addition *to the second Part of the* Essay towards a Natural History of the Earth.

THE *Confectaries* of the *former Part* of this *Discourse* are all *negative ; that* being only introductory, and serving but to free the Way to this *second Part* ; to rescue the Enquiry from the *Perplexities* that some *Undertakers* have *incumber'd* it withal ; and to set aside the false *Lights* they used in Quest of the *Agent* which transposed these *Sea-Shells* to *Land.*

Now, the only *sure Lights* we have to conduct us in the *ascertaining* this Affair, are History of Fact, and *Observations.* So that I shall give here some Intimation of the *Chief* of *those* that serve to clear up *this Subject,* and bring the
Thing

Thing in Queftion to a fair *Decifion.*
Thefe are, That the *Earth,* all round the
Globe, appears, wherever it is laid open,
to be *wholly compofed of Strata,* lying
on each other, in Form of fo many *Sedi-
ments* fallen down fucceffively from *Water.* That, accordingly, thofe *Strata*
that lye *deepeft,* are ordinarily the *thickeft:* and thofe that lye *above,* gradually
thinner, quite up to the Surface. That
there are *Sea-Shells,* and *Teeth* and
Bones of Fifhes, found *repofited* in thefe
feveral *Strata*; not only in the more
lax, *Chalk, Clay,* and *Marle,* but even
the moft folid *Stone,* and the reft. That
thefe *marine Bodies* are *incorporated*
with the *Sand* that conftitutes the *Stone*
of thefe *Strata,* in fuch Sort as together
to *compofe one common Mafs.* That on
breaking up this Mafs, fo as to part the
Shell from the *Stone,* *this* is ever obferv'd to have receiv'd an *Impreffion* of
the *exterior Surface* of the *Shell,* fo exact as to fhew it had been *contiguous* and
apply'd to *all Parts* of the *Shell*; which
the *Stone could* not be capable of, had it
not been then in a *State of Solution,* the

Mat-

Matter whereof it confifts *loofe*, and
fucceptible of *Impreſſion*. That, upon
breaking the *Shells*, and examining the
Infides of them, they are found to contain in them *Stone*, commonly of the
fame Kind with that without, which the
Stratum is made up of, and *apply'd* as
exactly to the *Infides* of the *Shells*; fo as
to have taken the *Impreſſion*, and all the
Lineaments of them, after the Manner
of Matter *caft*, foft, or melted in a
Mould. That the *Shells* are as frequently *immers'd* in the *Subftance* of the *Mi-*
neral and *Metallic Nodules*, even the
moft firm and folid, *Flint*, *Spar*, *Py-*
ritæ, and the reft; the Matter of thefe
Nodules exhibiting the *Lineaments* and
Impreſſions of both the *Outfides* and *In-*
fides of the *Shells*, as truly as the *Stony*
Matter of the *Strata* does. That thefe
Marine Productions are thus *repofited* as
well in the *loweft Strata*, as in the *upper-*
moft ; at the *Bottoms* of the *deepeft Mines*,
as to the very *Tops* of the *higheft Moun-*
tains. That they are obferv'd in fome
Places in fuch *Multitudes*, as in Bulk and
Quantity, to *equal*, if not *exceed* the

Sand,

Sand, or other *terreſtrial Matter* of the *Strata.* That there are ordinarily digg'd up, amongſt the reſt, *Shells* that are of *foreign Origin* and *Extract ;* being *not the Product* of the *Neighbouring Seas,* but of *Seas* much *remote,* and at great *Diſtance.* Thus we here *in England* diſcover, frequently at great *Depths, Shells* of Fiſh, very numerous, and of different Kinds, that appear now living on the Coaſts of *Peru,* and other Parts of *America.* That there are likewiſe *diſcover'd* commonly at Land, and in the Bowels of the Earth, *Shells* that are not at this Day found *living* on any *Coaſts ;* being doubtleſs ſuch as naturally reſide and inhabit only in the *deepeſt* and moſt remote *Receſſes* of the *Main Ocean,* without ever now approaching near any Shore, or being conſequently ever ſeen. That, in *all Parts* of the *Earth,* as well in *Aſia, Africa,* and *America,* as in *Europe,* as well in *Countries* the moſt *Diſtant* from any *Seas,* as thoſe that lye *near* to them, the *Strata* are compil'd, and the *Marine Bodies* diſpos'd in them, every where after the very *ſame Method;*

and fo, as apparently to fhew *Things* were reduced into this *Method* in *all Countries*, at the *fame Time*, and by the *fame Means*. That there are alfo lodg'd in the *Strata*, *Bones*, *Teeth*, and other Parts of *Quadrupedes*, or *Land Animals*, and oftentimes of fuch as are *not Natives* of the *Country* in which they are thus found. Particularly here in *England* we dig up the *Tusks*, and the *Grinder-Teeth*, the *Bones*, yea, whole *Skeletons* of very great *Elephants;* and likewife incredible large *Horns* of the *Moofe Deer*, a Creature not known to be *now living* in any other Country excepting *America:* As alfo, fometimes Shells of *Tortoifes*, peculiar to the fame Country. That there are befides, repofited in *Stone*, and even in the firmeft and hardeft *Strata*, Leaves of various Kinds of *Vegetables:* and fometimes whole *Trees;* as alfo fuch *Fruits* as are durable, firm, and capable of being preferv'd, *e. gr. Nuts*, *Pine-Cones*, and the like. That, amongft the reft, there are difcover'd, under Ground, *Trees*, *Leaves*, and *Fruits* of *Vegetables*, in *Countries* where fuch do

not

not now fpontaneoufly *grow*. Nay, that
there are digg'd up *Trees* in great *Numbers,* and many of them very *large* in
fome *Northern Iflands,* in which there
are at this Day *growing* no *Trees* at all;
and where, by reafon of the great *Bleaknefs*
and *Cold* of thofe *Countries,* 'tis probable
none ever did, or could grow. That, of
all the various *Leaves* which I have yet
feen thus lodg'd in *Stone,* I have obferv'd none in any other State, nor
Fruits further advanc'd in *Growth,* and
towards *Maturity,* than they are wont
to be at the latter End of the *Spring Seafon **. That the fquamofe Covers of the
Germina or Buds. and the Shives or
Chaff of the *Juli Trees* and *Shrubs,*
that fall off in the Spring, and are found
in fo vaft Quantities in many Peat-Marfhes, apparently point forth the *fame
Seafon.* As do likewife the immenfe
Sholes of the *Ova* of *Fifhes,* fo frequent
in

---

* *When,* according to the *Mofaic Relation,* the
Water of the *Deluge* came forth, and put a *Stop* to
the *Growth* of both *Animals* and *Vegetables. Confer
Part* 3. *Sect.* 2. *Conf.* 5. and *Part* 6. towards the
End.

in the upper *Strata* of *Stone*. That the *Shells* of the *Young* of *Fiſh* of the *current Year*, wherever digg'd up, are of the *Size* and *Bigneſs* they are uſed to arrive to at *that Seaſon*. That of all the many *Flies* and *Inſects,* that I have yet ſeen incloſ'd in *Amber,* I have never obſerv'd any that were not of the *vernal Tribes* and *Kinds.*

THESE are the main *Obſervations* whereon I ground what I *offer* in this ſecond Part of the *Eſſay towards a Natural Hiſtory of the Earth.*

A Mine-

A

Mineral Dictionary;

O R

An alphabetical

INDEX

Of the Names of all Kinds of Fossils,
referring to the Pages of this Work,
wherein each is explain'd.

Bar-

INDEX of Things occasio-
nally *treated of* in *these* Papers.

F I N I S.

Printed in the United States
By Bookmasters